Urbanisation in the Island Pacific

Managing rapid urban growth presents significant challenges in the small independent countries of the Pacific islands.

Although they originated in colonial times, the towns and cities are now distinctively post-colonial, with economies, environments and social structures that reflect unique island characteristics. This transformation has stimulated new concerns, such as the causes and effects of pollution, the need for employment for landless migrants, the need for adequate and affordable housing and the financing of expanding urban services.

In this book the authors explore the diversity of the urban experience in the ten independent island states, focusing on strategies to secure long-term sustainable development. The first volume to provide a comprehensive analysis of Pacific island towns, it examines:

- the historical, social and economic basis of urbanisation;
- the demands on new structures of governance;
- indigenous perceptions and responses to urban life;
- the role of a range of agencies;
- the place of urban centres in national development, and as conduits of globalisation.

The book will be essential reading for academics and professionals involved in urban and regional planning, sustainable development and Pacific island studies.

John Connell is Professor of Geography and Head of the School of Geosciences at the University of Sydney. He has taught and worked on a range of development problems in the Pacific for more than twenty years and has published a series of books including *The Last Colonies* (1998), *Papua New Guinea: The Struggle for Development* (1997) and *New Caledonia or Kanaky* (1987).

John Lea is Associate Professor and Head of the Urban and Regional Planning Program at the University of Sydney. He has undertaken research in the Pacific region since 1975, from Tonga to Micronesia and the Torres Strait, and his previous books include *Housing in Third World Countries* (1979), *Yellowcake and Crocodiles* (1986) and *Tourism Development in the Third World* (1988).

Routledge Pacific Rim Geographies
Series Editors: John Connell, Lily Kong and John Lea

1 **Landscapes of Globalization**
Human geographies of economic change in the Philippines
Philip Kelly

2 **Sex Work in Southeast Asia**
The place of desire in a time of AIDS
Lisa Law

3 **Urbanisation in the Island Pacific**
Towards sustainable development
John Connell and John Lea

Urbanisation in the Island Pacific

Towards sustainable development

John Connell and John Lea

London and New York

First published 2002
by Routledge
2 Park Square, Milton Park, Abingdon, Oxon, OX14 4RN

Simultaneously published in the USA and Canada
by Routledge
711 Third Avenue, New York, NY 10017

Routledge is an imprint of the Taylor & Francis Group

First published in paperback 2011

Typeset in Garamond by
Keystroke, Jacaranda Lodge, Wolverhampton

British Library Cataloguing in Publication Data
A catalogue record for this book is available from the British Library

Library of Congress Cataloging in Publication Data
Connell, John
Urbanisation in the Island Pacific : towards sustainable development / John Connell and John Lea.
p. cm.
Includes bibliographical references and index.
1. Urbanisation—Pacific Area. 2. Cities and towns—Pacific Area—Growth. 3. Sustainable development—Pacific Area. I. Lea, John P. II. Title.

HT384 .P16 2002
307.76'099—dc21 2001048445

ISBN13: 978-0-415-24670-5 (hbk)
ISBN13: 978-0-415-51384-5 (pbk)

In memory of
David Lea 1934–2001
and
David Drakakis-Smith 1942–1999

Contents

Illustrations

Plates

Figures

Tables

Preface

The origins of this book began when we separately arrived on the shores of the Pacific in the 1970s and became immersed in the diversity of island development issues. More recently, we both participated as researchers in the Pacific 2010 project initiated by Rodney Cole at the National Centre for Development Studies (NCDS). Our contribution to the research was to analyse and evaluate problems of urban management in the region and led to a series of visits to Melanesia, Polynesia and Micronesia between 1992 and 1996. These followed earlier visits and subsequent visits followed.

Though there have been many studies of particular aspects of urban growth and development in the Pacific, there has never been an attempt hitherto to compare the processes and outcomes of urbanisation within the region's independent states. This is not surprising. They are small countries, distant from each other, possessing a wide range of development problems and distinctive orientations to different former colonial powers. More significantly, their urban areas have generally escaped detailed academic attention, unlike the burgeoning cities of Asia, and data are unusually fragmented and difficult to obtain. Most data and most studies have emerged from Melanesia – which has three-quarters of the towns in the Pacific – hence, inevitably, this book has tended to have a Melanesian focus. Seeking to develop a comparative perspective on contemporary urban change in the Pacific has been a costly, sometimes frustrating, but necessary exercise. We have made the attempt because a baseline was needed to mark the advent of what some have termed the 'Pacific Century', and to examine the notion of sustainable urban development in a region that is in rapid transition. One example of that was the change in name of Western Samoa to Samoa in 1998, hence both names are used here according to context.

As the final draft was near completion, the rapidity and unpredictability of that transition became evident in a violent coup in Fiji in May 2000 that resulted in lost lives and a devastated economy, followed by a 'copycat coup' in Solomon Islands. Both were a response, in part, to ethnic tensions that were particularly evident in the capital cities, and to the problems of urbanisation itself. One outcome was migration – renewed emigration from Fiji and urban–rural migration in Solomon Islands – as a certain peace was slow to be restored in what was increasingly becoming referred to, by external observers, as one part of an 'arc of instability' from Aceh (Indonesia) to Fiji. Both of these disruptive political events in Melanesia have occurred at a time when instability and uncertainty appear entrenched in the most populous island state of all – Papua New Guinea. Necessarily this book can take little account of such recent, violent, yet localised impacts on urban life. It largely refers to rather more harmonious times of a year or so earlier, though it nevertheless primarily focuses on the difficult problems of management and development.

Acknowledgements

This book draws on three reports written by the authors between 1993 and 1997 on contemporary developments in the regions of Melanesia, Polynesia and Micronesia. We are particularly grateful to Professor Ron Duncan, Director of the National Centre for Development Studies at the Australian National University, for permission to use materials previously published as NCDS Pacific Policy Papers and to the University of Hawaii Press for permission to use material previously published in *The Contemporary Pacific*. Figure 1.1, drawn by Jane Eckelman, comes from the same source. Thanks are also due to Professor Rikki Kersten, former Director of the Research Institute for Asia and the Pacific at the University of Sydney, for permission to use parts of the report on urbanisation in Micronesia written for the Institute and published by the Centre for Pacific Islands Studies at the University of Hawaii in 1998. Particular thanks are extended to Rodney Cole, the former Director of the Island-Australia Program at the National Centre for Development Studies, for initiating the Pacific 2010 project that enabled formal work on Pacific urbanisation to begin in 1992 under the sponsorship of AusAID. We are grateful to Sandra Donnelly for her fine work deciphering our drafts.

Abbreviations

ADB	Asian Development Bank
AIDAB	Australian International Development Assistance Bureau (AusAID)
ANU	Australian National University
BLLPB	Betio Local Land Planning Board
BOOT	Build-Own-Operate-Transfer
CBOs	Community-based organisations
CFA	Compact of Free Association
CLPB	Central Land Planning Board
CRP	Comprehensive Reform Programme
DLSE	Samoa Department of Lands, Surveys and Environment
EEZ	Exclusive Economic Zone
FSM	Federated States of Micronesia
GDP	Gross domestic product
GNP	Gross national product
GTHS	Papua New Guinea Grassroots Toea-Savers Housing Scheme
HART	Fiji Housing and Relief Trust
HRF	Housing Refinancing Facility
IASER	Institute of Applied Social and Economic Research
ILO	International Labour Organisation
IMF	International Monetary Fund
IUIDP	Integrated Urban Infrastructure Development Programming
KALGOV	Kwajalein Atoll Local Government
KLIS	Kiribati Land Information System
KMHARD	Kiribati Ministry of Home Affairs and Rural Development
MIRAB	Migration, Remittance, Aid and Bureaucracy
NCD	National Capital District
NCDC	National Capital Development Commission
NCDS	National Centre for Development Studies
NGO	Non-governmental organisation
NHC	National Housing Commission; from 1990, National Housing Corporation
NLTB	Fiji Native Land Trust Board

PC	*Post Courier*
PIIDS	Pacific Islands Industrial Development Scheme
PNG	Papua New Guinea
PWD	Public Works Department
SAP	Structural Adjustment Programme
SPACHEE	South Pacific Committee for Human Ecology and Environment
SPARTECA	South Pacific Area Regional Trade and Economic Cooperation Agreement
SPC	South Pacific Commission (Secretariat of the Pacific Community)
SPREP	South Pacific Regional Environmental Programme
TFR	Total fertility rate
TPNG	*Times of Papua New Guinea*
TTPI	Trust Territory of the Pacific Islands
UNCHS	United Nations Centre for Human Settlements
UNDP	United Nations Development Programme
UPNG	University of Papua New Guinea
USP	University of the South Pacific
VULCAN	Vila Urban Land Corporation
WCED	World Commission on Environment and Development
WHO	World Health Organisation
WSTEC	Western Samoa Trust Estates Corporation

1 Introduction

An urban Pacific

> Our island world ceased to be. The world exploded and our island became a remote outpost . . . the last place in a country which has few centres and much remoteness.
>
> (Luana 1969)

The islands and seas of Oceania, from the vast thinly populated island of New Guinea to the tiny atolls of Micronesia and Polynesia, cover more than a third of the globe but have a population of just 7 million, of which rather more than half is in Papua New Guinea (Table 1.1). Despite the sparseness of its settlement, Oceania is exceptionally varied in language, culture and ethnicity with isolated islands and impenetrable mountain ranges emphasising difference and enhancing diversity. A quarter of the world's languages are spoken in the region reflecting extreme geographical and cultural variation, especially in Papua New Guinea (PNG), which alone has over 800 languages, and Vanuatu – with over a hundred – has more per capita than any other country. Despite its recency, colonial contact has had a powerful and wide-ranging impact, resulting in the initial decline of indigenous populations in some areas, the imposition of new languages, political and legal institutions and economic systems. Christian missionaries transformed belief systems, created new and often deep divisions – important in very small states – and emphasised old differences. In the nineteenth and twentieth centuries substantial new migration movements, such as that of Indian labour into Fiji and Filipino workers into Micronesia and beyond, have transformed demographic structures, accompanying crucial shifts in economic structures. Rural–urban migration and international migration (resulting in declining or stable populations in parts of Polynesia) have influenced identity, nationality and sovereignty. Urbanisation has transformed the once idyllic, but inaccurate, images of an unspoiled Pacific island paradise into something more mundane.

Though the countries and territories in the region are mostly small in population and land area (Table 1.1), they may be large in other respects. Exclusive economic zones (EEZs) have brought with them economic jurisdiction over vast areas of ocean and increased the economic and political significance

Table 1.1 Geographic, demographic and economic indicators of Pacific island countries

Subregion and country	*Land area (km^2)*	*Sea area (1,000 km^2)*	*2000*		*1998*		*1999 Annual (%) inflation*	*1998 Human Development Index (HDI)*
			Pop'n (1,000)	*Pop'n density (persons/km^2)*	*GDP (US$)*	*GNP per cap. (US$)*		
Melanesia								
Fiji	18,300	1,300	825	45	2,684	2,110	1.7	0.67
New Caledonia	18,600	1,740	213	11	—	—	—	—
PNG	462,243	3,120	4,791	10	1,196	—	16.0	0.31
Solomon Islands	28,370	1,500	420	16	926	750	8.0	0.37
Vanuatu	12,190	780	200	16	1,231	1,270	2.5	0.43
Micronesia								
FSM	701	2,500	118	168	2,070	1,880	—	0.57
Guam	541	—	148	274	—	—	—	—
Kiribati	810	3,600	91	112	702	1,180	2.0	0.52
Marshall Islands	181	1,942	52	286	1,882	1,540	1.0	0.56
Nauru	21	320	12	545	3,450	—	6.7	—
Northern Marianas	471	—	77	163	—	—	—	—
Palau	490	—	19	39	8,027	—	—	0.86
Polynesia								
American Samoa	200	—	64	321	—	—	—	—
Cook Islands	240	1,800	19	79	4,947	—	1.4	0.82
French Polynesia	3,520	5,000	233	66	—	—	—	—
Niue	259	390	2	7	3,714	—	—	0.77
Samoa	2,934	130	169	58	1,060	1,020	0.3	0.59
Tokelau	12	290	2	125	—	—	—	—
Tonga	650	700	100	154	1,868	1,690	4.4	0.65
Tuvalu	26	900	10	381	1,157	1,068	7.0	0.58
Wallis and Futuna	255	—	15	57	—	—	—	—

Sources: Secretariat of the Pacific Community (2001); UNDP (1999); World Bank (2000b)

of the region. Great distances can separate countries and internal island groups. In Kiribati it is more than 4,000 km from the capital in South Tarawa to the most remote Line Islands, and national boundaries straddle both the Equator and the International Date Line. Smallness, isolation and remoteness impose constraints on development, with few countries having markets large enough to allow significant import substitution and only Fiji possessing an industrial sector of any size. Consequently the demand for imports is high and there is a corresponding need to generate increased exports which are mainly limited to commodities. In certain countries, tourism and mining play an important role. Development strategies and prospects differ widely from country to country with the better prospects being in the larger countries of the south-west Pacific and the poorest in the small atoll states of the central Pacific.

The small scale of island economies creates considerable inflexibility in responding to changing global economic circumstances. In the atolls an extremely restricted physical environment has enabled only limited productive opportunities. A negligible share in commodity markets has led to a weak bargaining position and, in recent years, falling prices have weakened the economic position of most island countries, the effects being particularly severe in parts of Melanesia where the logging of old growth forests is fast depleting the environmental assets of future generations. Several countries earn substantial income from mineral exports but most national economies are based on agricultural production and fisheries. Fiji has a significant tourist industry but almost everywhere else isolation and high costs have largely held back growth in visitor numbers. The smallest countries are particularly aid-dependent, with most foreign aid in the South Pacific originating in Australia, New Zealand and the EEC and in Micronesia coming from the United States.

Political fission and fragmentation emphasise the economic problems of the Pacific islands, some of which are inherent in the distinctive characteristics of island states: limited natural and human resources, the many dimensions of isolation and fragmentation within countries (since none of the independent states, except Nauru, is a single island country). Constraints to development have made national economies fragile and vulnerable, maintaining cultural diversity and limiting the achievement of national identity and cohesion. Historic interests in the Pacific on behalf of the traditional colonial powers have given way to more demanding economic interventions by multinational traders and speculators.

Three externally constructed cultural and geographical divisions have been used to describe Oceania and have become entrenched over time (Figure 1.1). Melanesia is the largest and most complex region and extends from West Papua (a province of Indonesia) to Fiji, a country which shares many affinities with Polynesia in its hereditary chiefly systems. Melanesia is characterised by its many languages, cultures and physical and economic diversity. Colonial contact was more belated here than elsewhere in Oceania and in large parts of highland PNG it occurred as recently as the post-war years, one of the few parts of the world where 'first contacts' were captured on film and tape recorder. Not surprisingly indigenous cultural and belief systems play an exceptionally important role in everyday life. The two other regions of Polynesia and Micronesia are mainly composed of small islands, though size alone is not a direct measure of prosperity. In parts of Micronesia, especially in the Federated States of Micronesia (FSM), social and political organisation is still primarily based on hierarchical hereditary systems, which can be both feudal and caste-like as in Yap. In the somewhat larger Polynesian states social organisation is elaborate, often formal and with hereditary leadership still highly important. In Tonga the monarch has more authority and power than the recently introduced democratic system and in several Polynesian countries social stratification remains resilient, resistant to change and crucial for the nature and context of development. Despite the presence of modernity in various guises, the historic social and geographic divisions remain. Never rigid, however, they have blurred with population movement and the emergence of new identities.

The politics of Pacific development

Oceania was the last part of the world to be 'discovered', colonised, settled and decolonised, with just Tonga maintaining a tenuous independence in the colonial period. Only in the New Hebrides, as it became Vanuatu, was there violent transition to independence. Colonial history has been of enormous importance, not least because in the nineteenth century most of the tiny urban centres of the Pacific were simply places of colonial administration and trade. By 1906 the western powers (Great Britain, USA, Germany, the Netherlands, Spain and France) had finally completed the colonisation of Oceania. Colonial control was characterised by a succession of principal actors; Spain departed the region in 1906 and Germany lost its empire (most evident in Samoa, New Guinea and parts of Micronesia) after the First World War. Japan took over German Micronesia between the wars and the rest were shared by New Zealand (Western Samoa, Niue, Tokelau, Cook Islands) and Australia (New Guinea and the United Nations Trusteeships of Papua and Nauru). After the war the United States took over the Japanese colony of Micronesia on behalf of the United Nations, as the Trust Territory of the Pacific Islands.

Before the Second World War only Western Samoa, with its long history of resistance, had an independence movement. Elsewhere, fragmentation, limited social and economic development, and isolation prevented substantial challenges to colonialism. In 1962 Western Samoa became independent but elsewhere the decolonisation process was slow; independence was rarely actively sought, whilst colonial powers considered many colonies too small to be individually viable. Nauru[1] emerged from its period of trusteeship in 1968 and, between 1970 and 1980, most of the larger remaining colonies of consequence gained their independence: Fiji (1970), PNG (1975), Tuvalu and Solomon Islands (1978), Kiribati (1979) and Vanuatu (1980). More than half the entities in the region nevertheless remain politically dependent. Most are quite small and, other than in New Caledonia, French Polynesia and the Cook Islands, have no real urban centres. They have therefore been excluded here.

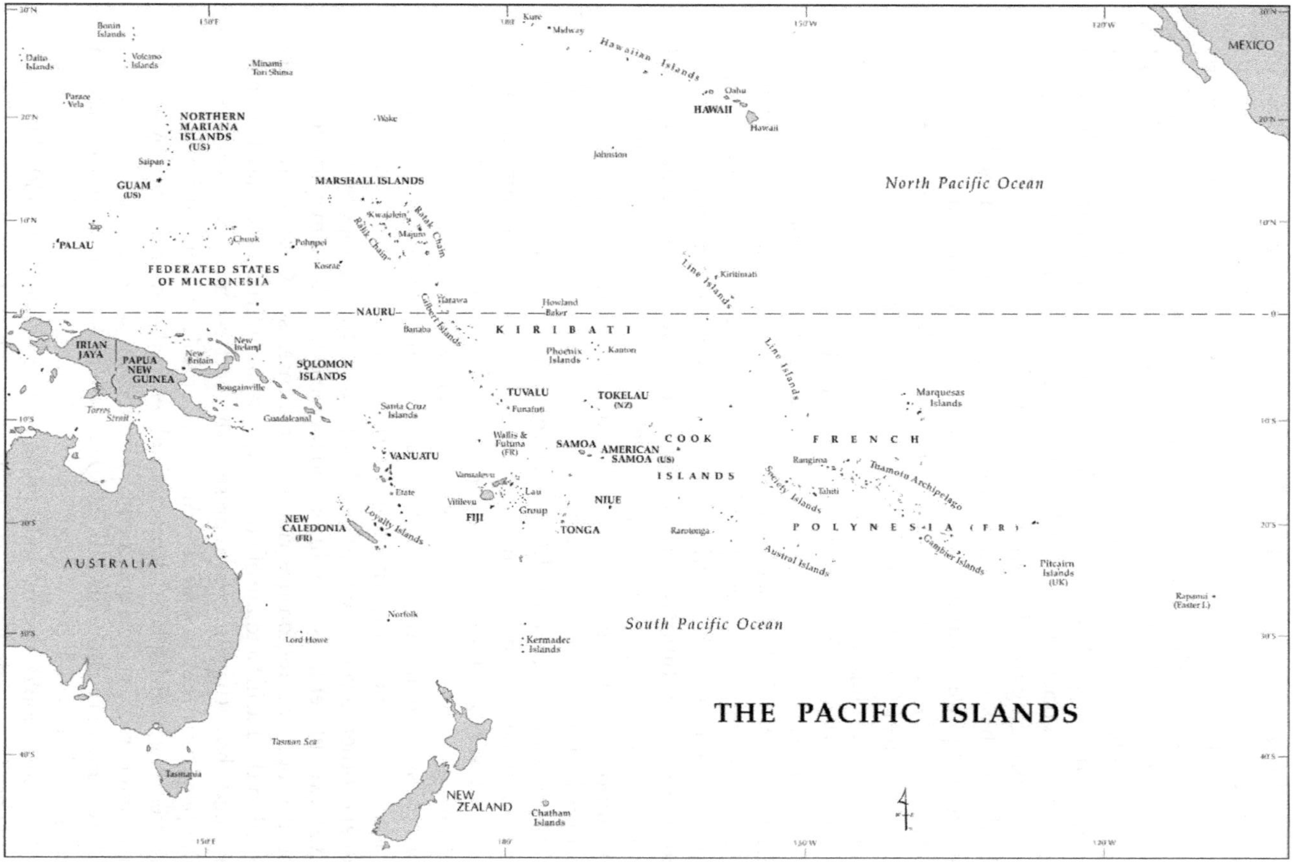

Figure 1.1 The Pacific islands

Source: *The Contemporary Pacific*, © University of Hawaii Press

The experience of the four Micronesian entities that comprised the former US-administered United Nations Trust Territory of the Pacific Islands (TTPI) has differed in several respects. In 1976 as the trusteeship moved towards its anticipated close, the Northern Marianas sought to increase links with the United States by becoming a Commonwealth. Over a period of several years, in the 1980s the Marshall Islands and the FSM negotiated Compacts of Free Association (CFA) with the United States, which gave them virtually full independence (with some restrictions on foreign policy) and very substantial medium-term aid packages. Palau (Belau), the fourth territory, failed to satisfactorily negotiate a CFA until 1993 but became independent in 1994, apparently bringing to a close the era of decolonisation in Oceania.

Island states have generally adopted government systems modelled on those of the colonial powers. All have elected assemblies or parliaments but possess different electoral situations. Though parliamentary democracy prevails, entities such as the Great Council of Chiefs in Fiji exercise considerable power. By contrast the development of political parties has been limited and few philosophical or even practical differences exist between them. Everywhere parties are primarily associated with prominent individuals, many of whom are traditional leaders. In Melanesia especially, parties are numerous and characterised by shifting memberships within parliament, and governments are composed of equally fluid coalitions that are prone to disintegration. This seriously limits the ability of parties, and hence governments, to develop and implement long-term policies and programmes. Some political systems in the region are particularly complex, as in Palau, which has a bicameral national legislature and sixteen state governments, for a country of just 19,000 people. The FSM is the only federal political system; in most situations the four FSM state governments are more important than the national government. Most people have very little involvement in national or local politics (where this exists) and there are few effective political organisations to bring this about.

Political systems are still in transition in most places as parties emerge and fragment, yet several distinctive features colour much of the discourse about urban and other development issues. Political cultures are personalised and involve powerful elements of tradition and regionalism in a realm where loyalties to social group and region override national interests. Classes have scarcely emerged in the Pacific and are of no significance in politics. There are few trade unions and those that do exist represent the better-paid urban workers and usually play little part in politics. Armies exist only in Fiji and PNG, yet in both states they have displaced democratically elected prime ministers by violent means. Democracy is often qualified by ethnicity, made dramatically apparent in the two military coups in Fiji in 1987, where a Fijian army displaced a government dominated, but not led, by Indo-Fijians, and again, in 2000, when a 'civilian coup' toppled the first Indo-Fijian prime minister (Lal 2000). Ethnicity has also posed problems where a particular group or category claims sovereignty within a nation or, stopping short of claiming full sovereignty, demands a higher degree of legitimacy. Violence in the Solomon Islands in 1999–2000 followed accentuated tension between the islands of Malaita and Guadalcanal. Ethnicity remains of exceptional importance in Oceania at different scales and in different ways, especially in Melanesia where language diversity and the presence of many minorities leads to limited national unity. No country is more 'imagined' than PNG. In some cases certain ethnic categories have disproportionate power relative to their numbers because of fortuitous location close to an important urban centre such as the capital. The growing significance of this geographic power is evident in the cases of the Motu-Koitabu of Port Moresby, PNG and the Mele and Erakor village groups near Port Vila, Vanuatu, who effectively control access to critical areas of land on the urban periphery. The rising power of Malaitans in Honiara, despite their minimal access to land, was a critical element in the rise of ethnic tensions and violence there.

Colonisation of Oceania took little account of regional and ethnic diversity. Many ethnic groups were linked together under a single administration so that during the period of decolonisation both nationalism and regionalism emerged and marked attempts to return to traditional divisions. The dismantling of the United Nations TTPI and the secession of Polynesian Tuvalu from Micronesian Kiribati exemplified successful restructuring. Attempts at secession have otherwise failed, typified by the decade-long rebellion in Bougainville as it sought independence from PNG.

In most contexts women are disadvantaged. Throughout Oceania women may contest elections and vote, though this right was only gained in Western Samoa in 1990, but only exceptionally are women elected to political office and they are actively discouraged from doing so in several states. They have become everywhere more prominent in the workforce and there are active women's organisations throughout the region. At times of political and economic crisis women's rights have been ignored, they have lost jobs, been victimised and have experienced various forms of violence (see, for example, Lateef 1990). Sexual and physical abuse of women has increased since independence, notably in PNG, where domestic violence is unusually prevalent, especially in towns, cultural values discourage discussion of these issues and patriarchal judicial and police systems prevent prosecutions.

Population and migration

Pacific populations tended to be either widely dispersed over large rugged islands or scattered in small island settlements at considerable distances from metropolitan centres of commerce and power. The transformations in society and economy that accompanied western and Japanese intrusion into the area caused major changes to the growth and distribution of populations, stimulating further economic change, a degree of diversification and, above all, uneven development as societies and nations shifted from subsistence economies to cash-orientated ones.

With minor exceptions, and none in the eleven independent states, the populations of Pacific countries are now as large as they have ever been (see Chapter

3). By world standards the present rate of growth is high and for some analysts this has even suggested the prospect of a Malthusian crisis as population outstrips resources (Fairbairn 1983: 32). The post-war demographic pattern has been one of high but recently declining birth and death rates. This incomplete demographic transition has stimulated rapidly growing populations in many areas, especially in Melanesia and parts of Micronesia. In Polynesia natural increase has long been high but has been effectively siphoned off through emigration.

Over the past century the movement of people within and between islands has intensified in volume, increased in distance, and become more complex in pattern and purpose. In some countries and in parts of several countries migration is particularly significant. In the eastern islands of Fiji, for example, 'Changes in natural trends due to declining mortality, the impact of the family planning program and fertility decline from other causes, are overlain completely by migration' (Brookfield 1980: 185). Modern communications have intensified migration flows in a region which has historically been characterised by high mobility. International migration beyond the Pacific islands region is substantial in volume and impact, and small islands and remote areas are becoming depopulated. Increasingly each of these trends is combined in a fourth: the steady growth of urban populations marked by a reduction of circular migration and growing numbers of urban-born. International migration to foreign urban centres is primarily a Polynesian phenomenon with the principal destinations being New Zealand and the United States (alongside Canada and Australia), but Micronesians are now adopting similar strategies. This kind of movement has resulted in urban concentrations of Pacific islanders on a scale that is scarcely paralleled within the Pacific itself. Auckland has become the largest Polynesian city. Emigration has had a significant effect not just on the migrants but on their attitudes and contribution to development in their home countries.

The urban scene

Despite the significance of population movements, the actual extent of urbanisation measured in terms of the proportion of the population living in urban areas is limited. None the less there are more than thirty-five towns and cities in the Pacific with populations greater than 5,000 (Table 1.2), two-thirds of them in PNG and Fiji, though only the capital cities are growing rapidly. In Melanesia urban growth rates have scarcely increased since the 1970s and in PNG, at least, they have actually been falling. This was accentuated, brutally and dramatically, by the complete destruction of the town of Rabaul by volcanic eruptions in 1994, by the abandonment of the three Bougainville towns, with their population of over 20,000, during the civil war, and by the movement out of Honiara in 2000. In Micronesia too urban growth rates appear to have slowed even though town populations are proportionately very high. For Polynesia urban boundary problems prevent clear conclusions from being drawn. Towns

Table 1.2 Larger urban centres of the Island Pacific

Country	*Town*	*Population*[a]	*2000 est.*
PNG (1990)	Port Moresby	194,295	245,000
	Lae	80,655	120,000
	Madang	21,057	30,000
	Wewak	23,224	30,000
	Mt Hagen	18,977	25,000
	Goroka	18,368	25,000
	Rabaul	18,186	
	Daru	8,490	11,000
	Kimbe	8,363	13,000
	Vanimo	7,908	11,000
	Popondetta	7,627	10,000
	Bulolo	7,421	9,000
	Kavieng	6,486	9,000
	Alotau	6,435	—
	Mendi	6,192	—
	Lorengau	6,804	—
	Tabubil	4,670	5,500
Fiji (1996)	Suva	167,975	180,000
	Lautoka	43,274	45,000
	Lami	18,928	20,000
	Labasa	24,095	26,000
	Nadi	30,884	32,000
	Nausori	21,617	24,000
	Ba	14,716	16,000
	Sigatoka	7,862	9,000
	Vatukoula	7,079	8,000
Vanuatu (1999)	Port Vila	30,139	34,000
	Luganville	11,360	12,000
Marshall Islands (1988)	Majuro	19,664	25,000
	Ebeye	9,311	9,000
Samoa (1991)	Apia	34,126	36,000
Solomon Islands (1999)	Honiara	49,107	50,000
Tonga (1986)	Nuku'alofa	21,383	32,000
Kiribati (1995)	South Tarawa	28,350	32,000
Palau (1986)	Koror	9,442	12,000
FSM (1994)	Kolonia	6,660	8,000
Tuvalu (1991)	Funafuti	3,839	4,500

Sources: Connell and Lea (1993, 1995, 1997); Secretariat of the Pacific Community (2001)

Note

a The population data refer to the most recent census date (under country). Other than for Tuvalu, the table only includes towns with populations greater than 5,000. Other than in Tuvalu, Palau, Vanuatu and Samoa, there are several smaller towns unlisted here. Rabaul was destroyed by volcanic eruption in 1994. The population of Honiara fell from the above figures by about half in 2000, following ethnic tensions. In 2000 the new town of Nasinu was created in the middle of the Suva–Nausori corridor.

and cities nevertheless continue to grow, not only in the larger countries but also in smaller ones, such as Kiribati and Tuvalu, where they pose distinct social, economic and environmental problems.

It is no longer possible to regard Pacific islanders as solely rural people; only in Melanesia (and the very smallest colonies) are less than a quarter of the population living in urban areas. There were few towns in the Pacific before the twentieth century and even as late as 1960 only Suva and Noumea had populations of over 25,000. Yet in the 1960s urban growth accelerated. For Melanesia, in what were still colonial times:

> The rapid growth of towns equals in significance any of the changes that have swept across Melanesia since World War II. In its consequences for the integration of colonies into new nations, it may even be the most far-reaching of all changes, yet these places are still essentially communities of migrants.
>
> (Brookfield with Hart 1971: 384)

Urbanism remained in its infancy.

This rapid urban expansion followed the initial post-war and subsequent post-independence growth in government activity and spending, leading to a boom in bureaucratic jobs for the educated elite and those with skills, followed by increases in service employment. The speed of urban growth since the 1960s emphasised the importance of migration, increased densities in urban areas, reduced opportunities for urban agriculture, either in gardens or at the town boundary, and contributed to a movement away from sustainable urbanisation.

Throughout the independent states of the Pacific urban centres have transformed national landscapes. Indeed, in small states like Palau and the Marshall Islands, where more than two-thirds of all residents live in the towns (see Chapter 3, Table 3.1), they have become the new landscape. Most towns began as administrative centres, mainly in the late nineteenth century, and quickly acquired commercial functions; many smaller towns have never developed beyond this basic dichotomy. Capital cities, especially the largest, acquired a diversity of functions, a cosmopolitan population and sometimes acute social and administrative problems. A cluster of smaller towns were primarily company mining towns; some, like Bulolo (PNG), have now largely lost this function whilst Arawa (Bougainville) – formerly the largest mining town in the Pacific – and nearby Panguna were abandoned during the civil war. An even larger PNG town, Rabaul, remains a ghost town after its volcanic destruction. Such places provide more than metaphors for what could still be the transience of urban life, especially in Melanesia, where neither PNG, the Solomon Islands or Vanuatu have more than a fifth of their populations living in town, and where many are first-generation residents. Despite their small size relative to the towns of most developing countries, Pacific urban centres exhibit considerable complexity and diversity (Figure 1.2).

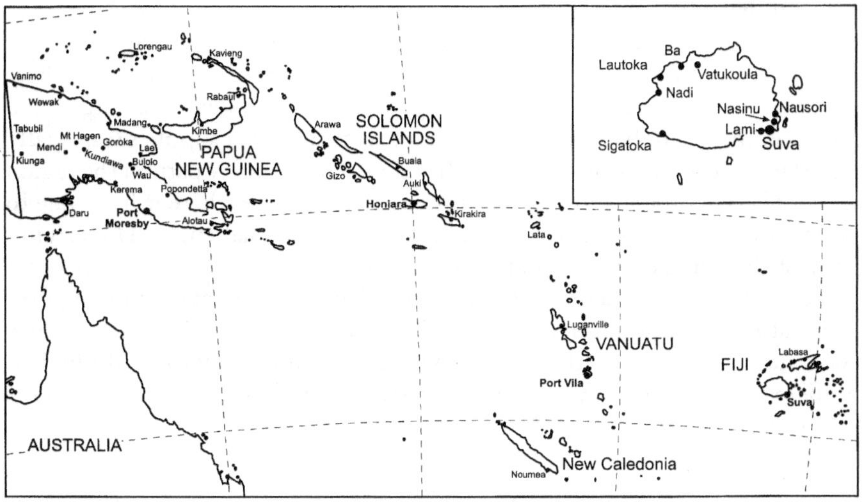

Figure 1.2 Melanesian urban places

Contemporary urban growth has been accompanied by various development problems and growing national concern. These initially focused on rising unemployment, the growth of unserviced and unsightly shantytowns, increasing crime rates and environmental degradation. Such fears were eloquently addressed by the first Prime Minister of PNG, Michael Somare, on the eve of independence:

> Do we really want to become a country of big cities? In all the 700 languages of our country we have never needed words for slum, for unemployment, for air pollution. Do we really wish to build the kind of country that needs those words?
>
> (Somare 1973: 2)

Not only are these words necessary but others such as crime, gangs, begging and poverty are now also familiar, as they are in some other Pacific urban centres. Tensions between land owners and migrants in the face of land shortages, bureaucratic ineptitude and even corruption have all contributed to disarray and division. Yet urban centres remain places of opportunity – social, economic and educational – and sites of both cultural change and conflict.

In recent years environmental aspects of urbanisation have received more attention, as housing no longer meets the rigorous standards of colonial times, traffic problems worsen and waterways and open spaces become polluted by the wastes of consumer society (Bryant 1993). In extreme cases such as in Tarawa lagoon (Kiribati) and Chuuk lagoon (FSM) the disposal of sewage that has polluted shellfish resulted in severe outbreaks of cholera at the end of the 1970s.

In the late 1990s and 2000 new outbreaks occurred at Pohnpei (FSM) and Ebeye (Marshall Islands). Development problems in the Pacific have quickly come to resemble those in other developing countries. So too have responses, or the lack of them. A weak national economic base has placed urban managers (and there are few of them) in a desperate race to cope with uncontrolled squatting and to mount an increasingly expensive search for water and fuel supplies (especially in the largest Pacific island city, Port Moresby). There are few parts of the region where effective urban land use planning exists, let alone where it has come to terms with meeting long-term social and ecological challenges. Underlying many of the approaches to urban growth in the region is a deep concern that the Pacific way of life is a rural way and that urbanisation, or at least its negative connotations, is a foreign way. Urban management and planning are also particularly difficult in the Pacific where economic growth is slight and municipal government, even when it exists, is starved of funds, yet the Pacific has an urban future.

Towards sustainable urbanisation

In the past decade Pacific countries have experienced political and constitutional crises as institutions inherited from colonial times struggle to adapt in the post-independence era, economies stagnate and poverty and environmental degradation have become more visible and more widespread. As the centre of gravity of national populations has shifted toward the capital cities, environmental problems, unemployment and social stress have all worsened. Ethnic, regional and cultural differences have fuelled political instability and uncertainty. The 1990s and now the advent of the new millennium have been particularly volatile – coups in Fiji, riots in Vanuatu, sustained ethnic violence in the Solomon Islands and the bid for secession in Bougainville – with most issues being related to minority problems, concerns over land tenure and employment and questions of identity and development. Few have been effectively resolved and many are accentuated in towns and cities, where the poor and dispossessed concentrate, as inadequate bureaucracies take on the difficult task of urban administration.

The unsustainability of present urban trends affects all aspects of policy formulation in Pacific towns and creates dilemmas over how the most robust and enduring features of island society can be reconciled with the unwanted outcomes of western economic models of development. There is no universally accepted definition of sustainability, but the concept has been of considerable and increasing significance, in a variety of environmental ideologies and practical situations. Numerous definitions exist, such as that of Bartelmus (1992: 15), where sustainable development is 'the set of development programs that meet the targets of human needs satisfaction without violating long-term natural resource capacities, and the standards of environmental quality and social equity'. Underlying this definition are three general themes: equity, both between people in the present (intra-generational equity) and between them and future generations (inter-generational equity); the ability of the environment to continue supporting the community at a certain level that does not decline through time; and the need for an adequate quality of life and the satisfaction of human needs. This involves economic and social development over a long period (Thistlethwaite and Davis 1996: 2–3) and takes sustainability beyond the realm of environmental management alone. In the Pacific this has resulted in an emerging focus on people's participation in, and responsibilities for, their own development (South Pacific Commission 1992; United Nations Development Programme 1994; 1997), issues that have become increasingly important in any assessment of sustainable development.

Much of the debate over sustainable development within and beyond the Pacific has focused on rural development (and natural resource depletion) alongside environmental management, and has scarcely been related to the urban situation. The goals of sustainable development as applied to cities largely derive from the World Commission on Environment and Development (WCED, the Brundtland Commission) definition of sustainable development, that it is necessary to meet 'the needs of the present generation without compromising the ability of future generations to meet their own needs' (WCED 1987: 8). Mitlin and Satterthwaite (1996: 31–2) suggest that the first part of the definition, 'the needs of the present generation', forms the development component of sustainable urban development and involves three kinds of needs: economic needs, which include access to an adequate livelihood and economic security when unable to work; social, cultural and health needs, covering the right to adequate serviced housing, good health and a home environment protected from hazards and pollution (with the special needs of children and child rearers – mainly women – commonly indicating a maldistribution of income within and between countries that requires correction); and political needs, which include the freedom to participate fully in decisions that govern urban and neighbourhood management. The second part of the definition, on needs of future generations, means ensuring that natural capital of all kinds is not depleted. This necessitates minimising the use and waste of non-renewable resources by lessening dependence on fossil fuels in all forms of domestic and economic activity and substituting them with renewable alternatives where possible. Waste of scarce resources can be minimised through reduced use, reuse and recycling, protecting irreplaceable cultural assets and controlling the extent of the urban 'footprint': the regional impact of urban growth and development (Rees 1992; Drakakis-Smith 1995; Mitlin and Satterthwaite 1996; McIlwaine 1997). Movement towards the achievement of these sustainable development goals must take into account the particular context of urbanisation in Oceania.

This book seeks to trace the context and history of contemporary urban development in the Pacific, and the manner in which this has moved towards or away from sustainable development. Development problems of the kind illustrated here may not necessarily be overcome by individual cities or nations alone but may require national, regional and international initiatives. Global warming, just one key environmental issue threatening urban sustainability and development, especially in low-lying island states (Connell and Lea 1992), originated outside the region but has critical local impacts. Other issues are more local, bound up, for example, in questions of land tenure. Thus the scope of the book and the debate includes: first, a review of how urban settlement became established in the region; second, identifying the factors underlying urban growth and change; third, understanding the distinctive features of the urban economy; fourth, examining the institutional experience of urban management; and fifth, looking at the more important sectors of urban policy, such as housing, land use and utility services, that support urban life, and the social context in which these operate.

At the start of the third millennium and for the first time in history, more people in the world live in urban rather than rural areas. Even in the small island states of Oceania some countries – such as Palau and the Marshall Islands – are already there. But urban planning and services delivery may remain inadequate without a much greater understanding of the needs and problems of urban residents, and especially the poor. These inadequacies are becoming more visible and costly and less amenable to solution, here as elsewhere (Roth 1987; Wegelin 1990; Gilbert 1992). The key subsectors of urban infrastructure examined here are those commonly defined in comparative policy studies, namely water supply, environmental sanitation services and housing. Integrally associated with these services are land management and the status of physical planning. The adequate provision of such basic services is an essential prerequisite to sustained 'economic, physical and cultural development in an accelerated and equitable manner' (Dimitriou 1991: 196).

Issues associated with urbanisation are likely to be of increasing importance in the years to come. Population projections for several countries suggest that population growth, particularly in the Solomon Islands, the Marshall Islands and Vanuatu is likely to remain at a high level until some time into the next century (Pirie 1994). In much of Micronesia and Polynesia, however, migration alongside population stability or even decline might be a more plausible scenario (Ward 1989; see Fry 1997). Simultaneously, continued low agricultural prices have reduced the economic incentives for many to remain in the rural sector and rural–urban migration, for this and other reasons, has been of continued significance. Most states (especially Fiji) are moving towards economic policies that, sometimes by default, encourage the concentration of economic growth in urban centres. Such policies place increased pressures on urban infrastructure which was constructed in colonial times for much smaller towns and cities; in the post-colonial period some of this has fallen into disrepair, as problems of recurrent funding, maintenance and the delivery of urban services have made the necessary 'good housekeeping' and extension of the existing system difficult. Standards that proved adequate in the past are less likely to be sufficient in the future as expectations increase and urban living becomes more permanent and accepted.

Throughout the developing world urbanisation is creating new opportunities for economic and social development but at the same time cities are experiencing serious problems of physical and economic management that endanger the hopes and the health of their growing populations (Annez and Friendly 1996: 12). New and determined efforts are required to ensure that sustainable urbanisation becomes a reality. In an appraisal of public services in PNG, the World Bank described a series of development problems including the rise in violent crime, increasing school drop-out rates, the extent of epidemic disease, and very limited and poorly maintained public facilities, but it also concluded: 'This situation does not have to persist. The quality of life can be significantly improved, as many of Papua New Guinea's Asian neighbours have demonstrated' (World Bank 1995b: 2). That challenge and opportunity remain to be answered there and elsewhere in the Pacific.

An urban Pacific

In an island region, the image of which is all too often that of a place of palm-fringed lagoons populated with rural people whose traditional cultures are intact and whose diet is based on local products, reality has become rather different. A steadily growing proportion of the population live in towns where economic, social and environmental problems abound. They are places whose experience to date has little to do with notions of sustainable urban development. This is to be expected because for much of the region and for most of colonial history towns were small, alien creations in which Pacific islanders were barely tolerated and sometimes actively discouraged (Chapter 2).

Continued high rates of population growth in most of the region have been accompanied by rural–urban migration, the effects of which are lessened in the case of the larger Polynesian states by relieving international migration (Chapter 3). As the economies of the islands evolved and diversified, independence brought greater urban employment opportunities – both in formal wage occupations and the informal sector – leading to rapid urban growth but more evident development problems (Chapter 4). In recent times a slowdown in the pace of economic growth has exacerbated urban unemployment and natural increase has generally become a more important influence on urban population growth than migration.

Urbanisation has become something of a paradox in countries where a rural economy predominates, where national development plans stress rural endeavours and the ethos of life itself remains largely rural. This is particularly so since urban growth in the Pacific has become associated with poor, under-resourced and fragmented urban management which seeks to plan and control development, overcome housing difficulties, improve water supply and waste disposal, and settle land tenure disputes, all despite limited powers and restricted human and financial resources (Chapters 5–7). This book examines the evolution of the urban Pacific, its problems and successes, the complexity of urban lives (Chapter 8), and the challenges to achieving sustainable urban futures.

Note

1 Nauru is excluded from this book since it has no real urban centre, issues of 'urbanisation' have never been documented and the problems of development and urban and environmental management are quite different from those of other countries in the region (see McDaniel and Gowdy 2000). As contemporary parts of other much larger countries, West Papua and Hawaii are also excluded.

Part 1

The context of urbanisation in the Pacific

2 The colonial heritage

Few other regions of the world exhibit so little evidence of urbanisation in the pre-colonial era. Throughout the Pacific the population lived in villages or scattered hamlets, and there were very few central places of any kind where regional trading might have occurred. The most extensive trading networks, such as the *kula* ring of the Trobriand Islands (PNG) and the 'stone-age trade' of the New Guinea highlands, failed to result in central places (Hughes 1977; Allen 1982). There were no significant functional specialisations beyond village level, nor were there traditional state structures, as existed in southeast Asia, but a proliferation of tenuously linked small-scale societies. Even in Polynesia, where more complex hierarchical systems existed, there were no townships; elaborate ceremonial and burial grounds had no counterparts in daily life. Only the larger islands of Micronesia and Tonga provide evidence of urbanisation before European contact.

On Pohnpei construction of the ancient township of Nan Madol began around 900 years ago, and continued for over a century, as basalt was carved and transported to create artificial offshore islands and an elaborate city of fortresses, temples and an administrative centre, sometimes grandiosely referred to as 'the Venice of Micronesia'. It remained the capital of a royal dynasty until the eighteenth century. The old city of Lelu on Kosrae, with a later origin, was very similar in both form and function, and declined equally mysteriously (Fischer 1964). Neither left any legacy for contemporary times. It was not until the start of the nineteenth century that the first colonial trading posts became administrative centres in a few scattered coastal locations, mainly in Polynesia. Most were tiny, with little influence on their hinterlands, and primarily served the needs of colonial venturers. Only much later in the century was there any semblance of a form of urban development that would have been recognised in Europe, or even on Pacific margins. Tiny though such townships were, they were the basis of the contemporary urban system. Urban history was colonial history.

The early colonial era

European explorers found rural societies, most of which were highly localised, despite the complexity of some Polynesian chiefly systems. Nuku'alofa proved a rare exception to overwhelmingly rural socio-economic systems; when the first missionaries arrived in Tonga in 1826, it was already a chiefly centre. Otherwise the first real towns were colonial constructions that followed early trading and whaling expeditions, as demand arose for ports of call for victualling, trading and entertainment. Small townships sprung up, usually close to the sites of existing villages, where there were good harbours or anchorages. They were built around the nucleus of trading establishments, store sheds and the inevitable drinking houses, and later acquired churches and schools. The townships catered for some indigenous interests, providing employment and trading opportunities of various kinds, becoming the first centres for the diffusion of artefacts, ideologies and diseases.

Many of the earliest towns were in the central Pacific – Honolulu, Lahaina and Hilo (Hawaii) and Papeete in French Polynesia – but Apia (Samoa) and Levuka (Fiji) became townships in the first phase of urbanisation. Perhaps the first modern urban centre in the now independent Pacific was however again in Micronesia where Koror in Palau had already become 'an attractive little town' in the 1830s:

> A wide stone pier extending a quarter of a mile out to sea was joined to a quay of the same length that ran along the beach. . . . A broad road, paved with stone ran up from the harbor, while stone-lined paths criss-crossed the town in every direction.
>
> (Hezel 1983: 178)

Levuka became significant in the 1840s, initially as a shipbuilding centre. Apia developed at around the same time, with missionaries, whalers and traders all taking advantage of the excellent harbour (Ralston 1977: 55). Levuka, where mountains prevented expansion, eventually gave way to the more central and spacious site of Suva, which became the capital of Fiji in 1882. Like Levuka, not all towns prospered. Despite its excellent harbour, Koror was isolated from trade in the region and the first modern town declined during the nineteenth century.

The earliest port towns of the now independent Pacific states grew slowly. In 1860 Apia's population was estimated at about 120; even then it had a law and order problem and was known for a time as the 'Cairo of the Pacific' (Ralston 1977: 94–5). It was formally brought under development control by the Municipal Convention of 1879, whereby the three interested colonial powers, Germany, the United States and Great Britain, negotiated the establishment of a self-governing enclave on neutral territory. A Municipal Board, which raised revenue and looked after public works, was perhaps the first municipal authority in the Pacific. Apia still only numbered about 120 'persons' in 1880 but there was already cultural diversity, with thirty-one Americans, twenty-seven Germans, forty-three British, five Spanish or Portuguese, two French and twelve Chinese (Pringle 1989: 26), but undoubtedly Samoans swelled those numbers. There as elsewhere slow urban growth was not unusual in the second half of the

Plate 2.1 Levuka, Fiji. Levuka, the former capital of Fiji, is now a World Heritage site. © John Connell

century, but as the number of expatriates increased, so too did racial and social exclusiveness, marking a shift from the tolerance and egalitarianism of the very earliest periods of urban development to more formal European control and regulation.

Beyond Palau, urbanisation in Micronesia had modest beginnings and in the nineteenth century was largely confined to Kolonia (Pohnpei) and Jaluit (Marshall Islands). Kolonia, founded by the Spanish as Colonia de Santiago in 1887, was a small administrative centre in both Spanish and German colonial eras. The atoll of Jaluit, a major commercial centre in the second half of the century, by 1880 was the busiest port in Micronesia; the port-town, with just two substantial trading firms and a few modern houses was 'metropolitan indeed' compared with other parts of Micronesia (Hezel 1983: 290). In 1878 it became the centre of administration of the new German colony but did not grow until Japanese times. Elsewhere trade and an administrative presence were insufficient to support towns.

Other than in Fiji the urbanisation of Melanesia had yet to begin. Only at the end of the nineteenth century did the first tiny colonial trading posts become administrative centres in a few coastal locations. Madang (then Friedrich Wilhelmshafen) briefly became the capital of the German colony of New Guinea at the end of the 1890s, until the transfer of the capital to Kokopo and then Rabaul in 1910. There, in the first decade of the twentieth century, major disputes with local people over land alienation culminated in a revolt, the forced exile of the people of three villages and the confiscation of their land, a precursor of land issues that became more problematic in the twentieth century. Despite its centrality for German colonialism, Madang had a European population of just fourteen at the turn of the century; Chinese traders and indigenous workers did not swell that number substantially. Early European contacts in Papua led to Port Moresby being chosen as the site of British administration in 1885 – when just seven white people were there. By 1888 the process of segregating expatriate residential areas had already begun; even in that short time the physical form and the social institutions of the town, which lasted until after the Second World War, were in place. At the turn of the century:

> Port Moresby was a tiny, poverty-stricken outpost on the fringe of a vast and savage wilderness. . . . The town dwellers were mainly alcoholics, eccentrics or little men disguising their lack of social and intellectual stature behind a front of starched tropical drill . . . a community which then comprised only sixty-six permanent European residents living in squalid little iron-roofed houses on the stony hillsides about the harbour. The centre of social life in this isolated, malarial place was an appalling tavern. . . . [The] centre of commercial life was a tumbledown trade store given over mainly to the buying of sandalwood and the selling of cheap grog.
>
> (White 1965: 53)

Jaundiced though this perception might have been, it was indicative of a beachcomber town that had yet to contribute to local development.

Suva, by contrast, was in some respects the first planned town in the region, and 'an outstanding example of a town built exclusively at the dictates of a colonial government' (Ralston 1977: 216) to regulate administration, trade and the indigenous population. The original population were relocated to a 'new Suva', Suvavou, beyond the urban boundaries. Suva became the capital in 1882 and, with an excellent harbour, a large peninsula and a fertile hinterland, it grew more quickly than any other Pacific town of these times. Yet Suva was never large; in 1911 it had no more than about 6,000 people, though this was about 5 per cent of the national population and included many Fijians and Indo-Fijians, unlike the situation in New Guinea where the urban population were primarily expatriates. European residents in Suva sought to pass a European Reservation Ordinance in 1912, which set out regulations designed to exclude 'coloured residents from designated areas of the town'. Although there was no formal segregation, building regulations effectively guaranteed it, and social divisions were reinforced by separate library reading rooms, toilets and bathing places, and curfews restricted non-Europeans to their homes after nine at night (Mamak 1978: 26). Similar regulations were later established throughout Melanesia. In the New Hebrides (Vanuatu) there were hints of urban planning, after the area around Port Vila became the headquarters of various traders in the 1880s. In 1895 Port Vila was declared a municipality, Franceville, with administrative ties to Noumea in New Caledonia, but urban growth was slight. In the Solomon Islands it was wholly absent, as plantations were few and trading interests minimal, in often hostile environments.

By the end of the nineteenth century most of the larger towns had developed some modern characteristics. As they grew, they became characterised by racial divisions, which gradually took on class elements. Their economies were becoming more diversified as trading increased in complexity and steamships enabled better contact with the outside world. Indigenous houses gave way to more permanent constructions; foreigners moved into what became sizeable nuclei of predominantly white settlers, and 'certain areas became recognised as white preserves in which the [islanders] were visitors or employees (usually domestic) rather than neighbours' (Ralston 1977: 164). Political control intensified this marginalisation, emphasising uneven development. Yet urbanisation was of slight consequence in the early colonial years; economic opportunities were too few in a scatter of islands at the ends of empire.

The twentieth century

It was not until the twentieth century and the inter-war years that the economies of towns diversified beyond their trading and administrative functions. Predictably enough that process began in Fiji. Towns such as Lautoka and Labasa acquired sugar mills, as agricultural processing began, and mining towns later emerged there and in New Guinea. Urbanisation remained more characteristic of Polynesia, yet periods of colonial prosperity in the first half of the century brought urban development to various scattered parts of the region. Although in the German era Koror was no more than a government station, under the Japanese it again became a large settlement, with many modern buildings, factories, warehouses and houses, and several schools. From the small village of the 1920s, the town achieved a population of more than 3,000 by 1935. The Palau Branch and regional government of the Japanese mandated territories (Nan'yo Cho), located in Palau, provided for open spaces, saw that the streets were paved and adequately drained, and ensured houses were built with good foundations (Ehrlich 1984). At the end of the 1930s growth became even more rapid, when pearling began to yield substantial profits:

> During the winter months the presence of well over a thousand divers . . . fueled an economic boom, particularly in the central districts of the city, where most of the inns, restaurants, and entertainment businesses were located. . . . Bars, cafes, geisha houses, and brothels did a roaring business and the twang of the shamisen and the sounds of raucous laughter, singing, and rhythmic clapping carried over the calm waters of Koror bay.
>
> (Peattie 1988: 176)

Though not excluded by legislation, Micronesians did not work in the more skilled and influential sectors of the economy. The town had 8,500 people in 1939, when Palau's population was 28,000, almost twice its present level (Ono

2002). It was an urban scene quite different from most parts of the Pacific, where towns were smaller, had administrative functions, and cosmopolitanism was discouraged.

Hesitant urban growth elsewhere in Micronesia was primarily a result of intensive Japanese colonialism. By 1941 more than thirty firms had offices and warehouses in Kolonia, alongside schools, a hospital and government offices; the town had a population of about 3,000, mostly Japanese immigrants. Neither Yap, Chuuk or Kosrae had anything resembling urban centres until the last years of Japanese rule, when Colonia (alternatively known as Yap Town) became a minor settlement. Dublon, the administrative centre of Chuuk since German times, attracted more government buildings but the population probably never passed a thousand (until its pre-war militarisation). In 1941 it was 'a rustic, mean-looking little place where only the main street was paved and the side streets, intersecting the bedraggled row of stores along the shoreline were churned into bright glue-like mud after a heavy rain' (Peattie 1988: 211). In contrast to the urban centres of later years it was almost a model of urban design. Okinawan fishermen, traders and shopkeepers gradually became concentrated in Jaluit (Marshall Islands). By the end of the 1930s probably a thousand Japanese, alongside 2,000 Marshallese, created a substantial urban population (ibid.: 186). There, as elsewhere in Micronesia, the indigenous residents of towns were almost all from the immediate area. Towns were still colonial creations, largely occupied by migrants from the metropolitan power. Japanese colonial practice was not to exclude islanders from urban residence but in almost every other respect it sought to recreate the urban conditions found in mainland Japan.

Away from Micronesia urban growth was still slow, though between 1900 and 1926 Apia more than doubled its population, but the economic depression of the 1930s reversed the movement to town. By the beginning of the twentieth century Nuku'alofa was a small and thriving commercial centre and there were other less important settlements at Mu'a, Neiafu (Vava'u) and Pangai (Ha'apai). In the years before the First World War traders were as active in Neiafu as in Nuku'alofa and there was some small-scale manufacturing in both places (Walsh 1982a). There was little urban development in the inter-war years; the depression slowed commercial activity though rural–urban distinctions were beginning to emerge: 'the small population of Nuku'alofa was becoming increasingly sophisticated and familiar with foreign ways and technology but the means of diffusing this among the population at large even on Tongatapu, let alone the other islands, were all but non-existent' (Campbell 1992: 152). Simultaneously the Tongan regional centres became relatively less important.

Urban centres developed rapidly in Fiji in the inter-war years, especially in the sugar cane growing areas of Viti Levu, where Lautoka and Ba developed around sugar mills and market centres. Labasa grew, in a similar way, on Vanua Levu, whilst Sigatoka, Nadi and Nausori emerged as market centres. Even so, no more than 13,000 people lived in towns in the mid-1930s, still just 6 per cent of the national population. Suva, which had become a municipality in

1910, held 15,500 people in 1936, a substantial colonial capital, with twice the population of any other Pacific town (after Honolulu, Papeete and Garapan on Saipan). Moreover, despite Suva being seen by Europeans as a place of European settlement, it had become culturally diverse; Europeans represented only 14 per cent of the urban population; Indo-Fijians, Fijians, Rotumans and Chinese made up the bulk of the inhabitants (Chandra 1990: 159–60). Alongside Jaluit, it was becoming the first indigenous Pacific town, despite the numerous restrictions on local urban residents.

The first mining towns emerged in the inter-war years, at Wau and Bulolo (in the uplands of Morobe, PNG) and Vatukoula in Fiji, during the Pacific's second mining boom. By the 1930s Wau and Bulolo were established urban centres, regular flights linked them to the coast, timber milling had become a major ancillary industry in Bulolo (and later became the town's main function) and the growth of these towns – the most dynamic in New Guinea – was a major reason for the decision to site the new capital at Lae. Bulolo remained a company town, with acute divisions between company employees and many of those employed elsewhere, or unemployed, living in sub-standard accommodation with inadequate services (Fahey 1980). Wau was more obviously a mining town, rapidly growing in the 1930s, but largely stagnant in the post-war era, until by independence it was 'having something of a struggle to survive' (Jackson 1976b: 320). In Fiji, Vatukoula – created in the mid-1930s – was scarcely a town at all, lacking any semblance of planning, and with housing primarily confined to a few houses for managerial staff and barracks for the bulk of the workers, many of whom were from the distant Polynesian island of Rotuma. Urban infrastructure was minimal, and 'standards of housing, dining and sanitation at Vatukoula reflected the principles of stratification in the workplace, and its underlying racialist ideology' (Emberson-Bain 1994a: 112). Because of the single-sex nature of mining towns, and the parallel structures of employment and recreation, racial divisions were more entrenched there. The most negative elements of colonialism were epitomised in the work and residential regimes of these single-purpose towns.

In Melanesia, despite the rise of the New Guinea mining towns and the regional impact of mining, urban growth was trivial and of limited consequence. Port Moresby, despite being capital of the vast area of Papua, was still a tiny town in the 1920s. By the middle of the decade there were perhaps a thousand people there, more than half being Europeans and a smattering of Chinese traders. Beyond the urban boundaries were numerous villagers, many of whom worked in town; they did not figure in the census and Port Moresby remained 'the white man's place' (Inglis 1974: 47). At Burns Philp's store in Port Moresby the Europeans were fanned by *punkas* whilst Melanesian customers were served through a window opening on to the street outside. In 1920 an angry resident wrote to the *Papuan Courier*: 'Although we are in a native country, we have not the desire to be amongst native-inhabited buildings along the foreshore or within the town limits, and desire less to be continually annoyed by coloured folk permitted within the town vicinity' (quoted in Inglis 1974:

48–9). It was not until 1927 that the first sealed road was constructed, and no reticulated water system existed until 1941; it remained a 'sleepy, colonial backwater' for half a century with a population of barely 4,000 on the eve of the Second World War (Oram 1976: 32–62). Rabaul dominated urban development in what is now PNG, being the centre of the plantation industry, a major port and market, and, unusually, a town that was planned – with growth influenced by the original plans of the German governor Albert Hahl – though planning largely amounted to a gridiron layout and controls on Melanesian and Chinese residents. The dominance and prosperity of the town declined in the 1930s, with a slump in copra prices, and ended more dramatically with the eruption of Mount Vulcan in 1937, resulting in the evacuation of 10,000 people, the transfer of the capital to Lae, and uncertainty over its future (Johnson and Threlfall 1985). That uncertainty was compounded further by the almost complete destruction of what remained of the city in the Second World War.

It was not until an airstrip was constructed in 1927 that Lae formally became a town and a prototype for New Guinean towns built around airstrips (though most subsequent 'airstrip towns' were in the highlands): 'the airstrip in New Guinea is perhaps analogous to the railway station of an earlier era in America and Australia, because it has generally decided the shape and settlement pattern of the town' (Willis 1974: 83). Despite its crucial transport role, even in 1942, when it had briefly become the capital of New Guinea, Lae only had about 120 European residents, some sixty Chinese and perhaps several hundred Melanesians (ibid.: 84). It was actually a satellite of Salamaua, further south on the coast, which had previously become the administrative and commercial centre of the district and the port for the gold fields; it remained dominant until the war, when it was bombed out of existence, in common with some of the Japanese colonial towns in Micronesia.

Port Vila developed into a small and cosmopolitan township with a population of some 1,000 people during the 1930s, consisting of European settlers and administrators, from both France and Britain, Chinese and Japanese artisans and Vietnamese plantation labourers. In this early period the indigenous ni-Vanuatu faced direct controls preventing them from settling in town:

> Urban residence was highly segregated with each ethnic group living in specific niches and enclaves. Apart from this ghettoization, the residence of ni-Vanuatu in town during these years was severely regulated by the Colonial administration. Ni-Vanuatu who were not from Efate or were unemployed, could not remain for more than 15 days in Port Vila. After that, they faced deportation to their home island. But even Efatese and employed ni-Vanuatu from other islands faced discrimination in that they were not permitted to remain in town after 9 pm.
>
> (Haberkorn 1989: 7–8)

The Second World War stimulated urban development in the New Hebrides with the rapid introduction of urban infrastructure in and around Port Vila and

Luganville in Santo. Urbanisation was least evident in the Solomon Islands, where Tulagi, the administrative centre and chief port, had a population of barely 700, more than half of whom were Melanesians, on the eve of the Second World War.

The social conditions that characterised the Polynesian beach towns by the 1880s took several more decades before they were evident in Melanesia. The colonial administration was of no great significance until the 1920s, and prior to that, traders, itinerant miners and seamen had characterised urban settlements. As the economic composition changed, so the gender balance also changed, with white women gradually becoming more numerous (Inglis 1974). Racial divisions were always more acute than in Polynesia. From its earliest days Port Moresby was a segregated town. These were transient, insecure and impermanent communities, evident in the urban landscape. The towns of Melanesia were never beautiful, even where French influences were apparent in Port Vila or German influence in Rabaul. Some of the smallest, such as Samarai (Papua), were partly fashioned from houses literally transported from the declining Australian mining town of Cooktown (Nelson 1976). They were toeholds on what was still an unknown, incompletely explored region.

The earliest towns provided only the most rudimentary, unskilled employment to the indigenous population, who were actively discouraged from becoming urban residents. Bachelor wages, the provision of single men's quarters, compulsory repatriation after a maximum period of employment, restriction of occupational mobility and the denial of political rights, including the rights to organise and to strike, made it necessary for workers and their families to maintain a long-term dependence on the rural village economy. Regulations controlling the employment and mobility of wage labour, against the background of a colonial society based on a colour bar, profoundly affected the conditions under which the indigenous urban workforce was employed and accommodated. Control over the Chinese population was no less apparent. When Rabaul became the capital of German New Guinea in 1910, the township was largely built by Chinese, whilst one of the first Chinese in New Guinea, Ah Tam, was granted a thirty-year lease of 17 acres of land, which developed into a Chinatown (Wu 1982: 25, 29). Chinese urban residents were restricted to particular urban areas, and Chinese were sometimes forcibly displaced from white areas, as in Rabaul in 1925–6 when nine Chinese residences were pulled down and their owners 'induced' to move to the area set aside for them (Jackson 1976c: 422–3). All larger Melanesian towns had Chinatowns in the post-war years, one of the largest being in Honiara, followed by Rabaul, Port Moresby and Lae, and in each of these towns there was sharp social segregation between Europeans, Melanesians and Chinese. The economic distinctions were crude:

> The Europeans are the rulers, the teachers, the planters and the wholesale vendors; the Asiatics are the artisans and the retail vendors; and the Kanakas primitive agriculturalists, from whose midst are drawn the lowly laborer, the plantation hand, domestic servant and cheap sailor.
>
> (Lyng 1919: 69, quoted by Wu 1982: 33)

Such crude distinctions remained evident into the 1970s.

Elements of 'urban apartheid' were widespread (Griffin *et al.* 1979: 55; Connell and Curtain 1982: 463–4), and in these circumstances the ideology of an elite urban life style readily became established. In the rural areas migrants, whether in town or plantation, were recorded by patrol officers in the village book as 'absentees'; their true place was in the village where 'the social, the racial and the spatial were embodied in explicit linguistic and conceptual form' (King 1990: 9). Until 1966 Fijians required administrative permission to leave their villages. Women were particularly closely controlled. The Native Courts Code of 1927 made written permits obligatory and introduced controls on women's migration to such 'prohibited areas' as Suva and Levuka; women who remained in these areas without authorisation for more than one month became liable to arrest and repatriation (Emberson-Bain 1994b: 130). Until after the Second World War Melanesians were permitted in towns only under employment contracts, and such contracts were of limited duration; urban residence had a wholly economic rationale. In the 1920s Port Moresby residents of Melanesian origin were regulated by curfews; whistles or bells sounded at 8 a.m. and 9 p.m. to demarcate periods when they were allowed in town (Inglis 1974: 48–51; Connell and Curtain 1982: 462–5). The colonial towns were largely places where Europeans could work, live and enjoy the social amenities but where Melanesians had only a tenuous connection, via their workplace.

Polynesian towns were longer established, with smaller European populations and fewer controls over indigenous urban residence and life. None the less, as Albert Wendt's novels of Samoa vividly demonstrate, they shared critical characteristics with other Pacific towns. In the 1920s 'Apia looked like a vision one could touch, smell and taste but never quite possess' (Wendt 1977: 47) and its history was that of another world:

> it had been carted from across the seas and nailed on the shore with copra profits. A horde of settlers and transients – uncouth beachcombers, sailors, remittance men, and drifters from Europe and America – had come in search of adventure and women and quick profits. All these *papalagi* [Europeans] banded together and turned Apia into a bastion of *papalagi* arrogance and greed, into a small Europe, notorious for its bars and political intrigues. . . . The *papalagi* had come to stay in a town over which Samoans had no control but which ruled all Samoa.
>
> (Ibid.)

It was a place of difference: a vantage point into different ways of life:

> Apia was a place to be visited perhaps twice a year, marvelled at, and returned from with bewildering tales of mechanical marvels, ice cream, big stores that sold everything a person could want, and a picture theatre where cowboys (brave heroes) and Indians (villains) – in true stories, mind

> you – killed one another without any thought of Jehovah or of moral consequences.
>
> (Ibid.: 8; see also Connell 1993)

It was the place where *papalagi* influence was greatest, the source of innovations, discoveries and new ways of thinking about and imagining the world. Even so Apia remained distinct from rural areas, the centre of colonial administration and the symbol of a subordination that provoked a bitter struggle against the New Zealand administration. Even under a relatively liberal colonial administration, the town was an alien intrusion of still limited indigenous significance.

By the time of the Second World War urbanisation in the island Pacific remained in its infancy; though their impact had been felt in most parts of the region, the towns themselves were generally tiny. Apia in 1936 had some 5,300 people – including 1,000 'part-Samoans', some Melanesians, thirty-five Chinese and 260 Europeans; it was already very clearly a 'mixed-race' urban centre. Nuku'alofa had almost 4,000 people on the eve of the war, whilst Vatukoula – with 3,000 people – was as large as these two colonial capitals. Wau and Bulolo were not much smaller than Port Moresby. The significance of these mining towns was a measure of the exploitative nature of colonialism and economic development. Despite the expansion of Suva, the grandest of the southern colonial capitals, and Koror, its counterpart in the north, rural life and economies remained dominant.

Post-war urbanisation

While the Second World War was a brake on economic development in many parts of the Pacific, it provided a stimulus to post-war development, as colonial administrations sought to develop their neglected colonies. This was especially true in Melanesia, where towns gradually began to grow and take on more Melanesian characteristics. In Micronesia it was quite different; the Japanese towns were shattered and their considerable Japanese colonial occupants relocated to the mainland and Okinawa. Post-war reconstruction was belated and half-hearted. Koror was a shell of its former grandeur. By 1955, despite steady expansion, the population was no more than 2,200 and its attractions were listed as 'electric power, a movie theatre, a hospital, motor vehicles and the traditional prestige of Koror village' (Force 1960: 26). Even so, that represented about a third of the total population of Palau. The decline of Koror was well documented by an American traveller in the early 1960s:

> Changes indeed. We drove up the main street . . . The street had been solidly lined for two miles by government buildings, department stores, shops, studios, fine residences, radio towers, schools, steamship offices, bookstores, neon-lit cafés, Shinto shrines, Buddhist temples, Christian churches, theatres and geisha houses. Now it had largely reverted to jungle.

> . . . In this 'Paris of the South Seas' . . . there was no American housing better than converted sheet-iron quonsets or the equivalent.
>
> (Price 1966: 111–12, 130–1)

Belatedly in the 1960s a new phase of rapid urbanisation began in Micronesia, but not until some of the more impressive Pacific colonial towns of the century had crumbled into disarray. The shattered urban centre on Jaluit was never rebuilt; nor was Salamaua in Papua or tiny Tulagi in the Solomon Islands. New urban centres, Majuro, Lae and Honiara respectively, were eventually constructed to replace them. Rabaul and Lae were equally devastated; South Tarawa, where twenty Europeans had administered the Gilbert and Ellice Islands Colony, was in even worse shape. In Micronesia and Melanesia, colonial endeavours in the post-war years centred on reconstruction.

In the vast new colonial region – the Trust Territory of the Pacific Islands (TTPI), administered by the United States for the United Nations – there were some attempts to develop policies on urban development. Despite the considerable difference between towns on atolls and high islands, American attitudes to urbanisation were similar throughout Micronesia. It was initially recommended that urbanisation be discouraged, as remained the case in Melanesia (Oliver 1951: 5). In the next decade urban growth was non-existent, to the extent that policies urging a greater concentration of the dispersed population gradually became more acceptable. The principal policy document on economic development in Micronesia in the 1960s, the Nathan Report, argued that urbanisation would benefit economic and social development and

> facilitate the creation of a Micronesian entity to replace the present somewhat artificial association of a dozen or so somewhat similar but nevertheless distinctly different cultural, political and economic entities. Increased mobility can speed the replacement of local particularism with a cohesive Micronesia.
>
> (Nathan Associates 1966: 100)

Hence the report proposed the resettlement of outer islanders, and others in remote places, into the main district centres, where they might gain formal employment and contribute to the development of the economy. Although these recommendations were largely ignored, urban development tended to follow the pattern suggested in the report as public services became highly centralised within each district.

Despite the official encouragement of population centralisation and urban growth, the towns themselves were unimpressive, bereft of amenities compared with colonial times, and with inadequate infrastructure for their inhabitants. As one critic noted:

> The district centers serve as the towns of Micronesia, but the word should be used with caution, for they are nothing like towns in any developed part

> of the world. They are more rural villages than urban centers and most of them are crazy collections of little buildings made of concrete or sheet metal or packing cases or thatch strung at odd and individualistic angles, along winding, muddy, pot-holed dirt roads that climb hills and plunge into gullies and twist and turn as if their makers were drinking the fermented juice of the coconut palm as they worked.
>
> (Nevin 1977: 141)

In these depressing urban environments 'the rusting automobiles, decrepit toilets and ubiquitous beer cans symbolize the contact culture life style of Micronesia's district centers' (Marshall 1979a: 6). These unflattering accounts of Micronesian towns were indicative of the scale of wartime destruction and minimal urban infrastructure that independent governments eventually inherited from the American colonial era.

In the post-war years urban growth became more rapid throughout the Pacific. In Tonga trading activities outside Nuku'alofa never fully recovered and the primacy of the town became consolidated through increased government centralisation, the expansion of urban services and the development of some manufacturing industry. Its population grew to about 16,000 in 1964 (Walsh 1964: 40). Apia also grew steadily, reaching almost 21,700 in 1961, the year before independence, when it held 20 per cent of the national population, a similar urban proportion to that in Tonga. In Fiji too urban populations grew, from barely 25,000 in 1946 to over 60,000 a decade later, a jump to 17 per cent of the national population (though the main reason for growth was the extension of the boundary of Suva from 1 square mile to 6 square miles). From 1956 to 1966 there was even more rapid growth to a total urban population of 160,000 – a third of the population – again, partly through boundary extension, but also from the growing maturity of an increasingly urban economy. In each of these three states, where the war had little impact, towns were firmly established and urban populations more permanent than elsewhere. Economic development was more evident and communications better established.

Unlike most other Pacific towns, Suva's population was already becoming permanent by the 1960s. About 17 per cent of Fijian households in 1959 had been resident for more than 25 years; some had lived there for over 40 years, and most intended to stay, despite diverse links with what were still home villages (Ward 1959; Nayacakalou 1963: 34–5; Chao 1980). Even residents of squatter settlements, already significant by the 1970s, and of the mining town of Vatukoula were long-established and, in the latter case, had sometimes lived there all their lives (Ward 1961: 265; Whitelaw 1967: 7; Samy 1977; Connell 1985a: 80). By the 1970s more than twice as many Indo-Fijians as Fijians owned houses in towns, where they were more likely to be engaged in commercial pursuits (Nair 1980). None the less Fijian migrants from the smaller eastern islands, where employment and income-earning opportunities were poorest, and Polynesian migrants from outlying Rotuma were just as likely to be effectively permanent residents.

By contrast, the most distinctive characteristic of urbanisation elsewhere in Melanesia was its recency, most evident in the Solomon Islands, where it was not until the early 1950s that the construction of Honiara began; in 1959 fully 82 per cent of the town's population (then merely 2,800) were male, an acute measure of the novelty of urban life (Bellam 1970). Whilst the New Hebrides escaped war damage, urbanisation was no more rapid than elsewhere in Melanesia. The war brought thousands of indigenous ni-Vanuatu into contact with urban life for the first time; some stayed as restrictions on urban residence gradually eased. By 1955 there were still only 200 ni-Vanuatu in Port Vila (and 170 in Luganville), but neither town had more than 2,000 people; rapid growth brought their populations to 15,000 and 7,350 respectively by 1970. Even in PNG the situation was no different. Despite a generous definition of urban centre, the overall urban population was small. After the war, Port Moresby had a population of only about 5,000, but when it became the capital of the unified Papua and New Guinea in 1949, it grew more rapidly (Oram 1976: 84–8). In the 1966 census, Port Moresby had grown to 42,000 whilst Lae had 16,000 people; they were well on the way to becoming the most dominant cities in the country, whilst Port Moresby was soon to pass Suva, to become the largest city in the region.

If urbanisation was belated in Melanesia, it was even more belated in Kiribati and Tuvalu. South Tarawa was established as the administrative centre early in the twentieth century but was no more than a small colonial outpost for the first half of the century. In 1947, when a decision to establish the headquarters of the Gilbert and Ellice Islands Colony was made, the villages of South Tarawa had a population of 1,643 but they were scarcely urban. The period between 1963 and 1968, when the urban population grew to 10,616, marked the most rapid period of increase for South Tarawa, yet by 1978 it had reached 17,921, some 32 per cent of the population of Kiribati. Rather like the culturally similar Marshall Islands to the north, migration was particularly important in the 1960s but of reduced significance in the 1970s as the urban population consolidated. The early wave of rural to urban movements in the 1960s provoked considerable concern and, within the South Pacific region, Kiribati was one of the first countries to attempt to translate that concern into practical policies for regional development.

In neighbouring Marshall Islands equally rapid urbanisation ultimately posed similar but even more extreme problems. Until the 1960s urban migration was slight; in 1963 Majuro (Darrit-Uliga-Dalap) held just 3,774 people and Kwajalein (Ebeye) 2,388, but a decade later the respective totals were 8,003 and 5,469, following massive migration in search of urban employment; the urban population became 54 per cent of the national total, a percentage comparable only to that of Palau, where even by 1954 some 39 per cent of the national population lived in Koror, and in 1973 the proportion was 76 per cent, by far the highest in the region. Urbanisation was occurring without American encouragement. In FSM, alone in Micronesia, urbanisation was relatively slow, and little more than a quarter of the population were in urban centres in 1973.

Such rapid urban growth created critical urban problems in several parts of Micronesia, many intractable for a quarter of a century.

Throughout the Pacific, the 1960s and 1970s constituted the period of most rapid urbanisation. Where growth was particularly rapid, as in the Marshall Islands and Kiribati, new social and economic problems became evident, accentuating those of earlier decades. Racial discrimination and segregation were still evident in Melanesian towns. European, 'Asiatic' and 'native' quarters were differentiated for what were stated to be hygienic reasons; many 'public' institutions, such as parks, swimming pools and cinemas, were segregated and non-Europeans were denied access to certain jobs. Throughout Melanesia many Chinese migrated to Australia on the eve of independence or soon afterwards, and Chinatowns declined in size and significance. Discrimination against Melanesians was even more routine and pervasive, from store sale practices, to school entry, employment (and salary scales) and the use of vagrancy laws against unwanted migrants. Segregation was enhanced by basic geography: where it was possible, Europeans occupied the hills – Namanula Hill in Rabaul, Tuaguba Hill in Port Moresby, the Lenggaki and Vavaea Heights in Honiara, and the slopes overlooking Apia. In Suva, the best residential location had long been the sloping hilltop of Toorak, named after the fashionable Melbourne suburb, whilst more elevated Tamavua Heights followed as an elite suburb. Where hills were absent, European residential areas were in prime coastal locations, but segregation was less evident where topography was muted.

Segregation was not merely between Europeans, Asians and indigenous populations, but was also significant between indigenous groups, most evident in Melanesian towns, where there were sometimes acute divisions. In New Guinean towns like Lae and Rabaul there was a 'Papuan Camp', highlanders usually occupied distinctive areas in coastal towns, as did coastal migrants in highland towns – and there were often tensions between Papuans and New Guineans. In Honiara, Gilbertese residents were highly localised, as were migrants from Polynesian outliers such as Rennell, Bellona and Ontong Java. In Kolonia too, Polynesian outlier populations occupied distinct areas, and there and in Colonia (Yap) migrants from outer islands were marginalised in densely populated, poorly serviced compounds. Outer Islanders in Yap State were concentrated at Madrich, a tiny site of just 0.7 acres, in overcrowded barrack-style accommodation:

> Few people find Madrich an attractive place to live. It has been labeled a shantytown or a squatter settlement. Many of its structures are in disrepair and flimsily built. Little open space exists and sanitation is poor. The community has primarily depended on over-water privies and limited bathing facilities.
>
> (Alkire 1993: 52)

Active discrimination against Outer Islanders made urban life particularly difficult. Ethnic divisions characterised Pacific towns, though, in many contexts,

Plate 2.2 Madrich, Colonia, Yap, Federated States of Micronesia. All the occupants of Madrich are from the Outer Islands of Yap State, for whom access to other housing and land on Yap is difficult. Madrich, on an exposed site, was severely damaged by Typhoon Percy in 1990. © John Connell

these were overshadowed (and enhanced) by divisions between local landowners and all migrants from distant areas.

Social segregation and topography combined into a colonial urban geography, especially in Melanesia, where towns developed dual forms: an expatriate area, on a formal grid of land holdings and streets, with buildings little different from those of the metropolitan country, which 'coexisted uneasily' with indigenous villages:

> The dualism which thus developed was paralleled in administrative policy towards urban areas. The town was administered by expatriates, for expatriates and according to expatriate models. Public health, public order and the maintenance of so-called 'standards' required, it was thought, an element of insulation from the indigenous population.
>
> (Ward 1973: 366–7)

Suva, like other Melanesian capitals, thus contained distinct morphological divisions, urban villages (*koro*), effectively somewhat introspective compounds in contrast to the larger-scale, open layout of the expatriate city (O'Carroll 1997). Otherwise distinctions were emphasised in different ways and in different places by dress codes, restricted access to some 'public' buildings (including hotels) and so on, though these were more intangible in Micronesia and Polynesia. As the colonial era drew to a close the growing towns and cities remained obviously colonial creations, where archaic regulations lingered, and many indigenous peoples were made to feel unwelcome – if not always and only by Europeans. Yet in the post-war years more liberal colonial regimes, education and economic expansion stimulated rural–urban migration. Despite the existence of mining towns, there was little real economic diversity: towns were still primarily administrative and commercial centres, and the growth of employment – and formal housing – was inadequate to keep pace with the demands of growing urban populations. Well before independence social tensions alongside economic and environmental problems had emerged.

Towards independence

Throughout the Pacific, the 1970s proved to be a period of accelerated change, especially in political and economic terms. In several states, notably PNG, colonial powers sought to put in place the institutions appropriate to an independent state. Aid levels were particularly high, further stimulating urban bureaucracies. Such changes fostered minor construction booms, service sector expansion and new stimuli to urban growth. No longer were urban centres forbidden or alien to indigenous populations, but they were even seen as the 'crucibles of nationhood' (Ward 1970: 58), places where national identities might be forged in place of parochial ethnicities. Changes in education brought more high schools, and tertiary establishments, to the towns and cities; education more obviously geared graduates to urban life as political change was linked to new notions of modernity. At least as late as 1971 it was argued that 'urbanisation should be facilitated and encouraged' (Ward 1971: 81); then just four years before independence, only three PNG towns (Port Moresby, Lae and Rabaul) had populations of over 20,000 and the urban population was barely 11 per cent of the nation. Urbanisation in the Solomon Islands and Vanuatu was equally belated. During the late 1960s and early 1970s Port Vila grew at the rapid rate of 7.2 per cent per annum, boosted by changes in legislation which turned the country into a tax haven: 'The result was electrifying. Between 1972 and 1974 Port Vila became a boom town. Thirteen overseas banks opened their doors, the expatriate population tripled, subdivisions sprouted in what had been previously virgin bush and, as a side effect, tourism took off' (Forster 1980: 371, cited in Haberkorn 1989: 11). In the Solomon Islands only Honiara was an urban centre comparable to those elsewhere in Melanesia; its population had reached 15,000 in 1976, two years before independence.

In Tuvalu, urbanisation was one outcome of independence. Between 1973 and 1979, the census interval that covered the critical period of separation of the Gilbert Islands (which became Kiribati) from the Ellice Islands (Tuvalu), which occurred in 1975, Funafuti, the capital of the newly independent state, grew to 2,120 people. Many public servants, and others, returned from South Tarawa (now the capital of Kiribati alone). A year later, with a population of just under 1,500, Funafuti had 'on a small scale all the characteristics of a rapidly developing capital city in a developing country' (Howard 1976: 5). A few vehicles, a European presence, national government buildings, permanent housing, electricity, a large airstrip and conflicts and tensions over land gave Funafuti some distinct urban characteristics.

Elsewhere urbanisation was similarly rapid. The population of Apia doubled between 1951 and 1971, on either side of independence, but subsequently barely increased its share of the national population, as emigration replaced internal migration. Honiara doubled in size between 1976 and 1986, and Vila and Port Moresby grew almost as quickly over similar time periods (Connell and Lea 1993: 38–9). In Fiji, where the urban system was already well established, post-independence growth was less than in other parts of Melanesia. Only in Tonga, where the end of the protectorate brought no structural change, was urbanisation less rapid than in earlier years (Connell and Lea 1995: 32–3) and, as in Samoa, it slowed further with emigration. Both Nuku'alofa and Apia, to a greater extent than most other Pacific towns, have also been characterised by the growing significance of commuting from beyond the urban boundaries.

The concentration of increasingly indigenous political power in a single place stimulated the growth of national capitals, discouraged nascent decentralisation policies and, where urban systems did exist, as in Melanesia and Tonga, emphasised urban primacy – the dominance of a single urban centre – most evident in the smaller Melanesian states of Vanuatu and the Solomon Islands. Primacy was less apparent in PNG, partly because of the physical fragmentation of the country, but in Fiji the emerging Suva–Nausori corridor dominated urbanisation. By the mid-1970s Suva's population was four times that of the next largest city – Lautoka – and was growing faster: a situation rooted in the historical development of Fiji, with Suva the site of colonial power, government and trade. Policy decisions favoured Suva since the indigenous Fijian elite regarded south-east Viti Levu (including Suva) as the centre of their influence, and Lautoka was primarily an 'Indian town' (Chandra 1990: 165–6). More generally, public sector bureaucracies already dominated urban employment; more prestigious employment, 'formal' and 'informal', was there (including better opportunities for crime) alongside the best hospitals, while 'modern' social life (cinemas, restaurants, night clubs, organised sports) was sometimes absent in provincial towns. Primacy and centralisation discouraged the development of regional centres and made the provision of services in small towns more difficult (Ward 1990). Towns were primarily coastal, sometimes poorly located for national development (Jackson 1980), situations that did little to stimulate rural and regional development.

The most visible urban problem was the rise of squatter settlements. Illegal settlements, and thus significant socio-economic inequalities, were already apparent in Fiji by the end of the 1950s. In 1964 there were substantial areas of squatter housing, especially in the Suva suburbs of Raiwai and Samabula, and also Lautoka, to the extent that it was estimated that around one-fifth of the urban population lived in such housing (Whitelaw 1964: 54). In an attempt to cater for lower-income workers the colonial government set up a national Housing Authority in 1958 whose principal activity in its first six years of existence was to construct a 'low-cost' housing estate at Raiwaqa. The main problem experienced by the Housing Authority was that of bringing the cost of housing down to a point where the lowest paid workers could afford the rents; that problem was never resolved, and later proved unsolvable elsewhere in the Pacific (Chapter 7). There was no direct attempt to address the circumstances of those living in settlements. Indeed, by regarding Fijian urban settlements as 'villages' and Indian 'settlements' as being on an individual 'estate', they could be excluded from official concern and ignored in policy formation. Many squatter settlements exhibited a range of problems, from the absence of water and electricity to poor health conditions, the existence of 'dire poverty' and the necessity to 'fight for survival'; more than half the population of settlements lived there because they had been evicted by landlords, rather than because they had migrated from rural areas (Reddy 1976). In Fiji and subsequently elsewhere colonial and post-colonial governments largely chose to ignore squatter settlements, or persuade their occupants to return 'home', rather than devise policies to ameliorate housing conditions or the lack of services.

Extensive squatter settlement construction was more belated. Until the 1960s most migrant groups in Port Moresby had some relationship with Melanesians in the city or on its fringes, often based on traditional trading ties; but increasingly more distant migrants were establishing independent settlements in the area, mainly leasing land from local landowners. Settlement populations grew rapidly. In 1956 there were fourteen informal migrant settlements with a population of 1,850, but by 1970 there were forty with a population of over 12,000, almost a quarter of the urban population. At independence that proportion had increased to a third, and an acute division was emerging between the 'planned' and 'unplanned' sectors of the city, in the absence of any effective policies on housing and infrastructure provision. By the 1950s gangs were established in town, increasing in numbers into the 1970s, whilst crime too increased (Parry 1972; Po'o 1975), as urban living conditions became more difficult. Until 1970 the colonial administration maintained a policy of removing squatters from customary land, but it was ineffective. Policies aimed at providing lost-cost housing for Melanesians catered only for more affluent wage-earners, and some 85 per cent of the urban population could not afford the cheapest type of house provided by the Housing Commission (Oram 1976: 194). In 1973 a government White Paper, *Self-help Housing Settlements for Urban Areas*, for the first time argued that it was necessary 'to officially recognise the existence of as many of the settlements as possible and to establish basic services in them as rapidly as practicable' (quoted by Jackson 1976d: 53–4). The first 'site and service' schemes gave some assistance to those in settlements, but were only able to meet the increased demand for low-cost housing for a few years.

In the second largest PNG town of Lae, more than a third of the population lived in squatter settlements by the 1970s (Willis 1974: 142) and there were acute tensions between traditional village landowners within the city and the squatters. In Rabaul and Mount Hagen similar tensions prevented 'all but the hardiest of migrants from squatting on any but government land' (Jackson 1976d: 55). Sanitation and water supplies were often inadequate; in Wewak, wartime bomb craters were commonly used as bathing places and water collection points (ibid.: 61), and even in the smallest towns informal housing was perceived to be a problem. In the mid-1960s Wau Town Council sought to extend the urban boundary (of a town with a population of barely 1,000), demarcating a more distant boundary beyond which squatters would have to move (Jackson 1976b: 317–19). Responses to squatter settlements were rarely more subtle or enlightened, hence problems increased. By the time of independence great inequalities of wealth and opportunity in Port Moresby were replicated at different scales throughout the country. Whilst provision of housing and infrastructure in settlements was inadequate, and conflict over land made life difficult and insecure, communities were more socially balanced (in terms of sex ratio and family organisation) than some other parts of the city, and especially the more obviously transitory expatriate areas (Jackson 1976d: 57–61). In a sense, those who were most committed to urban life experienced some of the worst urban conditions.

In other Melanesian states the same problems recurred. In Honiara, in the decade before independence, there was a substantial increase in what were euphemistically, and perhaps optimistically, referred to as 'temporary housing areas', though some had been in existence for two decades. In 1979 some 15 per cent of the Honiara population – most originally from Malaita – lived in such houses (Oram 1980; Frazer 1981). As early as 1967, a decade before rapid urban growth, the official view was that there was incipient development 'of the sort of hideous and insanitary slum dwellings which are an unpleasant feature of so many African and Asian cities' (cited by Oram 1980: 139). Access to squatter settlements was tightly controlled. Anyone wishing to live in a squatter settlement had to apply for a temporary occupation licence, and had to be employed; the licence was valid for only twelve months (Kama 1979). None the less, concern grew over the impact of squatter development on soil erosion, the failure of settlers to pay rent, and health conditions in some settlements. The only policies directed towards squatter settlements were attempts to remove them (Connell 1983e: 51). As elsewhere, formal housing did not meet the needs of low-income urban residents and by 1979 fully half of those who needed housing could not afford the cheapest housing available for purchase, creating growing divisions within the urban housing market. Settlements were equally evident in Port Vila where, more than in other parts of Melanesia, migration came from disadvantaged, high population density rural areas. By the early 1960s it was apparent that many had inadequate sanitation and water services, although residents usually had legal status and security of tenure (Leaney 1966).

The most depressing urban conditions in the region were in Ebeye – and to some extent in Majuro – where exceptionally rapid urbanisation had followed the labour demand from construction of US military bases on Kwajalein, initially for the nuclear testing programme and subsequently as a missile range (which eventually became the 'star wars' programme). Micronesian migrants were concentrated on Ebeye atoll (with Americans on Kwajalein, an extreme and belated example of urban apartheid) at a population density which by the 1970s was one of the highest in the world. During the 1980s both Majuro and Ebeye continued to grow fast, with the population of the latter rising from 6,624 to 9,311 between 1980 and 1988, at which point it was considered 'supersaturated' (UNCHS and UNDP 1994: 107).

The population density on Kwajalein atoll had reached 570 per sq km by 1988. Since most people were on the island of Ebeye, just 0.2 sq km in area, the real density there was 46,000 per sq km and rising. Living conditions quickly deteriorated amidst the shacks that made up half the island's housing. Gangs emerged, crime and alcohol use increased, diets were often inadequate, and both Ebeye and Majuro had significant suicide rates (Alexander 1978: 155–69). Health problems on Ebeye were more serious than elsewhere in the country; in 1962 a virulent polio epidemic began on Ebeye and subsequently outbreaks of shigellosis resulted from overcrowding and the deterioration of

Plate 2.3 Majuro, Marshall Islands. © John Connell

sanitary and housing conditions (Storch *et al.* 1980). The Ebeye sewerage system never functioned properly (as a result of defective construction work) and at the end of 1981 the drinking water was 'one hundred per cent highly contaminated', such that 500 children had recently had gastroenteritis; Ebeye was a 'biological time bomb' (Topping 1980: 14), demonstrating in extreme form the general problems of rapidly growing, densely populated urban areas in less-developed countries.

Before independence land issues were critical in many Melanesian cities. In 1971 planning consultants observed that 'the success of the whole Lae plan will depend on the extent to which local right-holders can be persuaded to accept modern planning methods' (quoted in Jackson 1976a: 300). The presumptions and assumptions of such a statement were staggering, yet such presumptions were common throughout the colonial Pacific: with modernity and independence new values might simply replace the old, and land would become a commodity to be bought, sold and leased as in European cities. Nothing was further from reality. As early as 1964 villagers on the fringes of Maprik disputed ownership of part of the urban area, and prior to independence a group of villagers had demanded A$30,000 compensation from the government for the use of their land, or the removal of the town (Allen 1976: 352). Such demands were part of the early compensation rush in PNG (Chapter 6). Land shortages were apparent even in small towns, such as Kavieng and Vanimo, and urban problems were compounded by inadequate planning (even in the most basic sense of layout), the occupation of valuable sites by inappropriate functions (such as army barracks) and tensions between urban landowners and recent migrants, with insecure tenure, inadequate housing and poor access to services.

Land issues influenced housing provision, primarily because of questions of legality and insecurity of tenure, and the high densities of many settlements. Basic problems were exacerbated by low incomes and inadequate service provision. Kavieng, with not much more than 3,000 people, exhibited a range of problems:

> [nearby] villagers have indicated their determination to hold onto their land indefinitely. Expansion would be costly. Moreover the town's water supply is obtained from wells and an increase in population may endanger its stability. Tank water is inadequate in the dry season and refill water has to be carted from a limestone cave six kilometres away. Consequently toilet facilities are very poor. Most of the toilets in the low covenant areas are bucket latrines emptied daily by the sanitary truck; the smell can be very irritating.
>
> (Miskaram 1976: 413)

Urbanisation was fraught with unpleasantries and uncertainties. Even in the smallest Melanesian towns there were social problems, a function of inadequate housing and service provision, but also a result of ethnic divisions fuelled by gender imbalances and an absence of social amenities. In the tiny town of Buin on Bougainville, despite the population being no more than 750, Saturday afternoons were a 'time of tension' as restless, urban residents confronted each other (Connell 1976: 485). Conflicts between local and migrant populations were common throughout Melanesia and Micronesia, and emphasised the social, physical and economic problems of urban life.

Where urbanisation was particularly rapid, as in Micronesia, urban conditions were often the most difficult. By the late 1960s urban Tarawa had already experienced an increase in crime rates, problems of water supply, the erosion of the Green Belt in Betio and hence problems of local food production (and supply) and water pollution (Bedford 1968: 9–10). Child malnutrition was being combated by supplies of free milk powder, as the situation deteriorated further (Connell 1973: 403). By the end of the 1970s, as Kiribati went to independence, the social and economic problems identified at the start of the decade had all worsened. Unemployment was increasing, whilst the average household size in 1978 was 7.4 persons, compared with 5.6 in the outer islands, hence housing was more congested (Love 1982: 70–1). Growing urban land shortages, following land sales, emphasised the particular problems of the *kain Tarawa*, the indigenous occupants of the urban area (Takaio 1993). Illegal settlement on water reserve areas in Betio was presenting problems for water quality. Sanitation was critical because, although most households had formal facilities, sewage was traditionally disposed of by tidal action on the beaches. An outbreak of cholera in 1977 emphatically demonstrated the link between inadequate sanitation and health. Other amenities were also limited: a third of households used well water only and half cooked on wood fires. A mains electricity cable stretched the whole length of South Tarawa but access to power was too expensive for most households (Bailey 1984). Access to basic needs was exceptionally difficult for some urban residents.

The difficulties of urban living were fewer in the capital cities of Vanuatu and Solomon Islands than they were in Kiribati or PNG. Even so, in both Port Vila and Luganville, a range of problems was apparent, including marital breakdown, higher crime rates, excessive alcohol consumption and nutritional deterioration (Fox 1976: 74–5; Jabre *et al.* 1976: 4). Similar trends were occurring in Honiara (Connell 1983e: 50–1). Malnutrition had been widely reported in the region. In Fiji and Samoa the majority of cases came from poor rural areas, squatter settlements on the fringes of towns and low-cost housing areas, where low incomes prevented the purchase of adequate diets (Parkinson 1977), a situation that was also increasing elsewhere as populations became more sedentary, lacked access to land (and marine resources) and experienced unemployment (Coyne 1984: 109). Crime was evident in all urban centres, but especially the largest, notably in Port Moresby where, on the eve of independence:

> There seems to be a general reluctance on the part of both motorists and pedestrians to be about much after dark. . . . Car stealing has suddenly become a popular activity while house, school and store-breaking flourishes

> as never before. . . . Hardly a household in the city seems to have escaped the attention of the intruders.
>
> (Anon. 1974: 35)

Most of this increase in crime was attributed to juveniles, as in Suva (Chandra 1977), and the disruptive impact of migration, limited education, unemployment and social change (Gaigo 1970). While increased crime and other social problems clearly, and probably inevitably, accompanied accelerated urbanisation, a substantial part of the literature on crime and other issues represented the moral panic of expatriates fearful of a transition towards independence and local control over national development, convinced that the Pacific islands had rural futures and askance at the diminution of their own urban privileges. This was most evident in Melanesia; as one colonial official recorded: 'problems result from insufficient houses and accommodation, an absence of families, lack of services, limited recreational facilities, poor sanitary arrangements and a tendency to regard Solomon Islanders as temporary residents' (Allan 1957: 27). Colonial policy directions contributed to elite urbanisation, the lack of indigenous urban family life and an unwillingness to plan for urban futures.

As towns grew more quickly, there was more widespread colonial concern over the extent of urbanisation. In the Solomon Islands, urban growth was deemed inappropriate because of its impact on deculturation (Fox 1976: 33). In parts of Micronesia especially there was recognition that migration would hinder national development, as the rural labour force dwindled. As early as the mid-1950s in Palau, migration to Koror had left many rural states with populations heavily biased to the very old and very young (Force 1960: 26). The attractions of life in the more remote states were particularly limited: 'old and young eke out a subsistence living and receive cash handouts from relatives working in Koror and overseas. . . . Today the villages are virtually bankrupt' (McGrath 1970: 134). A similar situation was also true in remote islands, such as those of FSM and the eastern islands of Fiji, where populations were sometimes falling (Marshall 1979b; Bayliss-Smith *et al.* 1988). There was however necessarily some ambivalence; urban employers sought a stable workforce and favoured improved housing and other services that would encourage greater permanence, but in most places urban life was too unattractive and insecure to create an urban proletariat. Remittances flowed to rural areas and urban residents frequently expressed desires for retirement to home villages. Even in Fiji, where urbanisation seemed most permanent, few urban residents intended to remain there indefinitely, but the duration of stay was everywhere increasing, especially for migrants from areas where land shortages were evident (Fahey 1980; Morauta 1980). Even as Sir Colin Allan criticised administration policies, changing circumstances were apparent:

> Solomon Islanders have been looked upon by Europeans as non-permanent residents, who sooner or later will return to their homes and villages. A

> consistent shortage of housing and the high cost of living have discouraged Solomon Islanders from thinking of themselves as permanent residents . . . Many have shown signs of wanting to become permanent residents, and the demand will certainly increase.
>
> (Allan 1957: 125)

Towns would not always remain tiny, colonial creations and elite expatriate centres.

In the face of limited social change and acculturation, housing shortages and other problems such as increased crime and tension between urban migrants and established residents, particularly evident in Melanesia, few urban residents were likely to do other than express their rural hopes and dreams. Continued rural commitment contributed to urban anonymity and village-like residential patterns. Moreover many migrants could not afford the purchase of the land or housing that would provide a measure of commitment and saw little reason to do so, especially when urban economies experienced difficult times, as was apparent in Melanesia in the late 1970s: 'the world economic crisis has dissipated the mirage of easy money to be made in the towns; in such difficult times security comes from village and community values that are attached to group structures' (Bonnemaison 1985: 58–9). Some migrants even shared the administrations' perceptions of the tenuous nature of urban life: 'a person's proper place was back in the village. The fact that one was addicted to town life was nothing to be proud of' (Levine and Levine 1979: 1). Urban life was uncertain, alien and often discontinuous; migrants maintained ambivalence about urban attachments, preferring an 'ideology of independence' that stressed their ability to return to their rural 'homes' (Strathern 1975). Ironically, expatriates were even less likely to be committed to a long-term future in the Pacific towns, and imposed their own lack of commitment on those towns.

In several places policies emerged in direct opposition to urbanisation and, at least officially, in support of rural development. In the Marshall Islands the problems experienced by overcrowding on Ebeye prompted several direct attempts to persuade people to return to their home atolls from Ebeye, and in Kiribati policies favoured decentralisation and the settlement of the remote Line Islands (Chapter 3). In Melanesia there was often local indigenous opposition to rural–urban migration and urbanisation. On Aoba island (Vanuatu) leaders wished to prevent the long-term migration of men, because of its impact on weakening the community, although they did not oppose the movement of young men to work in more lucrative positions such as in the administration (Bonnemaison 1974: 274). In PNG too, urban migration was seen to be appropriate to elites, but not to 'ordinary' men (Strathern 1975) and, even less, to women (Connell 1985b: 65–7). Opposition to urbanisation was particularly evident amongst urban landowners, who observed and feared the loss of their privileged access to urban employment, the diminution of their land and social tensions. Even on Funafuti so rapid was urbanisation, and so great the tension between the long-established 'true owners' of Funafuti atoll and migrants from other atolls, that as early as the mid-1970s: 'Demands for restriction of entry to the capital were being expressed by women's committees and island councils in Tuvalu (the outer islands seeking to retain active young people; Funafuti seeking to retain its separate identity)' (Howard 1976: 25). In a remarkably short time, in the tiniest capital city in the world, urbanisation had challenged the old order. Elsewhere it was rarely welcomed by either colonial administrations or indigenous elites, new or old. In different ways the last years of the colonial era drew attention to increasing problems of urban development.

The colonial legacy

The beach communities of the nineteenth century and the administrative centres of the first half of the twentieth century were expatriate creations and expatriate centres. They were white men's (and infrequently white women's) towns. The towns were small-scale and peripheral colonial endeavours at the ends of empire: seedy and dusty outliers of distant and largely uninterested colonial powers. Nevertheless towns were 'administered by expatriate models. Public health, public order and the maintenance of so-called "standards" required, it was thought, an element of insulation from the indigenous population' (Ward 1973: 366–7). Different measures were devised by the colonial authorities to ensure the continuity of 'urban apartheid', a division that emphasised the European city in opposition to the indigenous village: 'the towns were toeholds of an alien society at the same time as they were bases for alien control' (Brookfield with Hart 1971: 390). Europeans were isolated by 'physical difference, social distance, ideological distance and by an overwhelming gulf of power and custom' (Worsley 1964: 33). Distinctiveness was regulated; assimilation, sexual relations and marriage between different groups discouraged, and multiculturalism wholly implausible. Land was firmly alienated where necessary. Colonial towns in the Pacific, whilst simultaneously offering some forms of economic and social development, established and regularised an ordered and largely intransigent inequality, above all in Melanesia, where race relations were worst.

The post-war growth in the public sector – and the service sector – resulted in elite European migration to most of the colonial towns, contributing to the emergence of new social divisions. In Apia, for example, a significant expatriate and half-caste (*afakasi*) population was further established in the town (Connell 1993), that gave some parts of it – particularly on the hill slopes – very distinct social characteristics. Whilst the larger cities, like Port Moresby and Honiara, had a characteristic form – with the most common feature being some relationship between ethnicity and altitude – most towns had escaped any semblance of planning. Indeed, in some states, such as Tonga – where Nuku'alofa was very much a collection of urban villages – other than a basic grid layout, planning was positively discouraged. Disorder was characteristic of urban management in these tiny and distant fragments of empire. Colonial administrations were more conscious of status and privilege than of development and equity.

In the widespread absence of land use plans and urban zoning outside the largest Melanesian centres, and because of limited modern development, at the end of the colonial era most towns and cities were more like clusters of urban villages than planned centres. There was minimal concentrated high-value development and high-rise developments were unknown. At the time of independence the only city with traffic lights was Suva (and some entire countries, including the Marshall Islands and Kiribati, are still without them), industrialisation was trivial everywhere, wages were poor and the costs of living increasing; yet urban populations were steadily growing. Housing problems were ubiquitous, most evident in capital cities, even where squatter settlements were located 'out of sight and out of mind' on marginal land, too distant, swampy or steep to attract formal housing construction. Colonial administrations preferred them not to be there, argued that Pacific islanders were rural people and lacked urban commitment, whilst stressing the virtues of rural development (but doing little to encourage it, beyond the plantation sector), simultaneously spending little on urban housing and planning (but promulgating an abundance of laws on public health and vagrancy issues). Despite the need for a committed urban workforce, wages and housing conditions were rarely adequate to produce commitment (other than in the few company towns); the more committed indigenous urban residents were often those who were trapped or dispossessed (Garnaut *et al.* 1977) in the sense that return to rural areas was difficult or impossible because either land was simply no longer available, or their claims to land had effectively lapsed over time. Colonial administrations found little incentive for substantial urban investment.

Pacific towns were none the less often classic colonial creations, of external origin and with a continued orientation, and dependence, outwards (symbolised by radio, and later television, channels). European and Chinese enterprises absolutely dominated the commercial scene; few large towns, other than Nuku'alofa, were without a Chinatown. Urban society was rigidly hierarchical; both bureaucracies and suburbs were structured accordingly, whilst most indigenous workers were unskilled (and many were male contract workers, detached from their extended families, and conscious of their marginal status). In PNG, Solomon Islands and Vanuatu urban wages rarely allowed significant savings or personal possessions, let alone a family residence (Connell and Curtain 1982). The indigenous population was unwelcome in most colonial towns: marginalised by legislation on mobility, housing policy and the structure of economic development, so much so that in the 1960s it was possible to observe of most towns that:

> people are separated into clearly demarcated categories; and there is a very high proportion of transients in the population . . . one gets the impression of a lack of corporate unity. One has the feeling that the town could pick up and move as the result of administrative fiat – as indeed Tulagi moved to Honiara. The town is artificial, with surface roots.
>
> (Belshaw 1963: 19–20)

Urban life was impermanent, uncertain and insecure, even when towns held out hopes of development and change; they offered new jobs and social services, including the highly desired secondary education, but such opportunities were few and the rewards limited.

Capital cities, like Honiara, were alien to most of the indigenous population, despite their 'parasitic impact' on the rest of the country (Bellam 1970: 93); this was even true of the smallest towns, such as the many district centres in PNG, that were unattractive to migrants, providing little employment, few services and minimal links to surrounding rural areas:

> They began as colonial stations to spread control and to bring communities into the cash economy, and they continue the same process. They are the representatives of an authoritarian administration attempting to make a nation of very diverse ethnic groups.
>
> (King 1983: 177)

Cities were places of racial tension, and in Melanesia especially there were indigenous tensions; urban social institutions, such as unions and welfare societies, were few (cf. Nayacakalou 1963; Oeser 1969) hence kinship ties, extending between rural and urban areas, were the most critical elements of social support. Few towns ever had municipal councils with any authority, power or resources. Narrow urban economies generated little of the tax base that could have contributed to urban government. Despite planning legislation, in some cases dating from the nineteenth century, the post-war towns were largely unplanned (especially in Micronesia, where urban planning regressed rather than progressed under American rule), services were never adequate for the more rapid growth of the 1960s and 1970s, and were overwhelmed by the growth of squatter settlements. There were significant regional variations; Melanesian towns such as Honiara, 'a classic example of the colonial city' (Bellam 1970: 94), were very different from such Polynesian capitals as Nuku'alofa, where colonialism was largely absent and where, despite the gridiron layout, the town remained more 'indigenous' than most (ibid.: 95) because of its underlying village structure. South Tarawa was very similar. Yet even the Polynesian towns, where expatriates were relatively few and indigenous authority considerable, remained places apart, outliers of a distant world of materialism and individualism, centres for the diffusion of decadence and development. Despite the flurry of colonial expenditure that usually preceded independence, urban problems usually worsened in those years, and raised uncomfortable problems for new national governments. The colonial urban legacy was sometimes insuperable.

3 Population growth and migration

Small populations, dispersed settlement and high mortality rates in Pacific island states have until quite recently combined to ensure little population growth. The Second World War raised expectations of better wages and living standards and, in the post-war years, colonial governments improved the basic infrastructure. However, in several places, including PNG (where a number of remote population groups were not contacted until the 1970s) and the FSM, basic facilities like primary schools and health posts were not provided in many districts until the 1960s and 1970s. The arrival of modern health facilities, alongside education and new sources of employment, hastened the pace of social change and the start of a demographic transition that, in turn, contributed to urbanisation.

Island populations

Outside Polynesia the region was much affected by the Second World War, which devastated some indigenous populations. Subsequently a new colonial order was established and services were restored and extended. Several innovations, notably DDT and penicillin, led to the reduction of malaria and the eradication of yaws, quickly improving health in Melanesia, where life expectancies were lower than elsewhere. In Polynesia populations were already growing; indeed a 1948 United Nations study implied that Samoa had the fastest growing population in the world. In Fiji, the Indo-Fijian half of the population exhibited both high fertility and low mortality, an unusual situation at the time. After a period of recovery from Japanese occupation, Micronesian populations – which had often declined in the inter-war years – also began to grow. Most Melanesian populations remained more or less static (Pirie 1994: 21). The impact of growing population pressure on national resources was first experienced in Polynesia. The transition to low mortality was ubiquitous, hence most island states became independent at about the same time that they were first experiencing exceptionally rapid population growth.

Throughout the Pacific life expectancies have risen over the past quarter of a century. Mortality rates are extremely low, though these are partly a function of a youthful age structure. In several states mortality rates by the end of the

1980s were five per thousand, but in the Melanesian states (other than Fiji) they were at least twice that level. Life expectancies are lower in the Melanesian states (with PNG having a life expectancy of no more than 54 years), whereas the Polynesian states have life expectancies in the upper sixties (Table 3.1). Infant mortality rates are rather similar in structure, with PNG and Kiribati again the worst in the region, and are even worsening in PNG. Most countries, but especially those of Melanesia, are only entering the early phases of the demographic transition.

Infectious diseases remain important, especially in Melanesia, where they are responsible for more than 20 per cent of all deaths. However, the health status of Pacific states is undergoing a transition; in most states non-communicable diseases are now more important causes of death than infectious diseases. This epidemiological transition towards 'diseases of modernisation', such as cardiovascular diseases and cancer, has imposed major burdens on urban health care resources since they are more costly to treat, posing a dilemma for resource allocation and reducing the availability of primary health care (UNCHS and UNDP 1994: 25–7). The incidence of HIV-AIDS has increased, especially in PNG, where it has reached potential epidemic proportions, and in Kiribati. Significant improvements in health status that occurred in the 1960s and 1970s are no longer evident. These trends have been exacerbated by urbanisation: a more sedentary life, combined with increased alcohol and tobacco consumption and poor nutrition, in contexts of inadequate education and housing, lack of access to clean water and sanitation, overcrowding and poverty (SPC 1994; Connell 1997b; Calvert-Faamoe *et al.* 1997). A major cause of illness among infants and children is gastrointestinal disease, particularly diarrhoea, which is also linked to inadequate water supplies and sanitation. Some of the worst health and mortality problems are experienced in urban settlements, especially in Port Moresby and the urban areas that have been established on coral atolls.

Fertility was low in most early Pacific populations, as many societies employed various techniques to restrict population growth; but controls over fertility weakened, with the growing authority of missions and administrations. The onset of the demographic transition in the Pacific has therefore been preceded by a period of rapidly increasing fertility. In the western Melanesian states that period is still underway; the Solomon Islands has a population growth rate of about 2.8 per cent and Vanuatu about 3.0 per cent. Though the population growth rate in PNG is only about 2.3 per cent, it is much higher in the more developed coastal areas, suggesting the future potential for more rapid growth. Growth rates are also high in Micronesia. In the Polynesian states and Fiji, the growth rates that caused alarm in the 1950s and 1960s have fallen, less because of lowered fertility than because of emigration, a 'safety valve' for continued high fertility rates.

Total fertility rates are very high in several states. In the Marshall Islands, Vanuatu and the Solomon Islands, they have only just fallen below six. In Polynesia they are somewhat lower, but with the exceptions of Fiji and Palau all the island states have rates above the global mean of 3.1. Fertility rates are

Table 3.1 Pacific populations

	Population	*Population density (persons/km^2)*	*Population growth rate*	*Urban population (%)*	*Total fertility rate*	*Life expectancy at birth*
FSM	118,100	168	1.9	30	4.9	65.7
Fiji	824,700	45	1.6	50	2.6	66.5
Kiribati	90,700	112	2.5	39	4.5	61.5
Marshall Islands	51,800	286	2.0	65	5.7	67.5
Palau	19,100	39	2.2	74	2.6	69.0
PNG	4,790,800	10	2.3	18	4.8	53.5
Samoa	169,200	58	0.6	24	4.5	68.4
Solomon Islands	447,900	16	2.8	18	5.7	61.4
Tonga	100,200	154	0.6	32	4.2	70.7
Tuvalu	9,900	381	0.9	44	3.4	67.0
Vanuatu	199,800	16	3.0	22	5.3	64.2

Source: Secretariat of the Pacific Community (2001)

high since contraceptive use is low, particularly compared with developing countries of similar income levels (Kane and Lucas 1986; Ahlburg 1996). Contraceptive use is particularly low in Micronesia; even in Kiribati, once regarded as something of a Pacific success story, 'many Kiribati women seem to be saddled with spending a good part of their reproductive lives pregnant as well as bearing the (health) consequences of frequent and closely spaced pregnancies' (Emberson-Bain 1995: 8). Fertility control is a taboo topic in public discussions in most countries. Throughout the region there is no indication that contraceptive use rates are anywhere above 30 per cent, and may be as low as 3–4 per cent in Vanuatu and the Solomon Islands. Many reasons explain the limited response to family planning, including the availability of children to perform work (especially in their parents' old age), children's potential to be income-earners, the desire and necessity to please husbands, the intrinsic need to produce and rear children (Nardi 1983) and limited access to family planning services. Urban fertility is lower than that in rural areas. Port Vila has a fertility rate of 4.5, compared with the national average of 5.3, and in 1994 more than half of urban women reported that they were current contraceptive users; those who were most likely to participate in family planning were better educated and more likely to be employed (or to have employed husbands), but also already had several children (Jayaraman 1995), a situation true elsewhere in Pacific towns. Even so these factors are only very slowly contributing to more extensive family planning.

The only country where there is relatively high access to family planning is Fiji, where in 1990 some 27 per cent of women were reported to be using contraception but usage amongst ethnic Fijian women was considerably below that of Indo-Fijians; however, in the mid-1970s almost 60 per cent of Fijian women were reported to be using contraceptives, and Fiji was held up as a model of success in the Pacific region (Connell 1987a). There, as elsewhere, family planning programmes have become weaker rather than stronger in recent decades, a result of government funding cuts, in a continued context of powerful religious objections and patriarchal societies in primarily rural societies. Throughout the region family planning programmes have largely been tokenistic.

The extent to which population pressure on resources has been perceived as a problem is extremely variable. Concerns were first raised amongst and about coral atoll populations. In Kiribati and Tuvalu this was apparent by the mid-nineteenth century; as early as 1865 it was suggested that the local people 'were the first genuine Malthusians. They feared that unless the population was kept down they would not have sufficient food' (cited by Munro and Bedford 1980: 3), and the people of Vaitupu atoll (Tuvalu) were reported to have such a fear of starvation that 'there was a rule that only two children should be reared in a family' (ibid.). By the 1890s resettlement was already being suggested as a solution to what was perceived as an impending overpopulation problem. In the 1940s, on a small scale resettlement had become a reality, with the migration of Vaitupu islanders to Kioa, a Fijian island purchased by the colonial administration (Koch 1978). Elsewhere in Polynesia there was little concern for growing populations, until well into the twentieth century; in several places numbers were growing only slowly in the twentieth century after sharp declines following colonial contact (see, for example, Maude 1981; Dening 1980). However, in the 1960s the United Nations drew attention to rapid population growth in Western Samoa, and both there and in Tonga the first significant international migration began. By the 1970s some in Tonga also perceived something of a Malthusian crisis. One prominent Tongan observed:

> If, as projected, our population doubles in the next 34 years, a very large part if not all, of our inland lagoon environment will perish. . . . The sands on our beaches are disappearing at what seems to be an alarming rate. . . . If our settlements grow physically they will swallow up much needed land for agriculture. . . . Our plans for future development must take into consideration the plight of our landless and our poor people who constitute a social phenomenon alien to our history.
>
> (Hau'ofa 1977: 3–14)

Though the rate of natural increase has barely changed since then, the predicted disasters were averted by international migration (and also urbanisation), as they were in Samoa, and population pressure on resources became more or less stabilised.

The early history of Micronesia, other than in Kiribati, was similar to that in Polynesia. In a number of places there had been population declines in the nineteenth century, growth did not generally occur until well into the post-Second World War period, and there was no perception of population growth being too rapid until much later. On the larger Melanesian islands, there was concern neither over population pressure nor of depopulation, other than in a few exceptional cases (Connell 1997a: 167; McArthur 1981). There was however considerable enthusiasm, especially among politicians, for more rapidly growing populations. In PNG, whilst there was official support for slowing population growth by the 1980s, this was rhetorical rather than real. When a tentative movement towards family planning was hinted at, the former Governor-General, Sir John Guise, wrote in 1987: 'You populate or you PERISH' (quoted in Connell 1997a: 173). This advocacy was generally accepted, though the problematic impact of rapid population increase had become apparent in some places. In Simbu, where population growth rates have not been high, 'it is only because the Chimbu have already sought their own solution to the problem by migrating to other areas that the pressure on existing areas is tolerable' (Howlett *et al.* 1976: 40). This situation was further emphasised in the post-independence years with the growing extent of 'functional landlessness' (Howlett 1973). The situation in Vanuatu and the Solomon Islands was much the same (Connell 1983b, 1983c). Throughout Melanesia indigenous responses to population growth are viewed in terms of migration rather than of direct controls on population growth, if the relationship is considered at all.

Although prolonged rural–urban migration, increasing pressure on both rural land and urban services, rising youth unemployment, social discontent and high levels of maternal mortality all contributed to a more favourable climate for establishing population policies, in PNG (McMurray 1992: 13) as elsewhere in Melanesia, achieving policy targets has proved impossible. The factors that are most conducive to successful population policies – integration of population and development policies, improved rural development and communications to spread new values and reduce the economic significance of children, formal sector employment opportunities for women, and increasing age of marriage (Ahlburg 1992) – are largely absent. The greatest probability of contraceptive use is amongst those urban women who are younger, educated, with high family incomes and some power (Agyei 1988: 107–10; Jayaraman 1995). Such women are relatively few. Consequently, whilst traditional controls on population growth have weakened, modern family planning techniques have scarcely begun to intervene. Growing populations are rather less evident in the Polynesian states because of international migration. Yet in some places, including Tuvalu, the possibility of migration has actually encouraged high rates of natural increase; at the household level, on Nanumea atoll, 'parents actively hope to produce remittance earners and most feel that this necessitates having more than one son [one of whom] will be our road to money and imported goods' (Chambers 1986: 283–4). Despite failures of family planning, the most rapid phase of population growth in Oceania is probably now over, even though growth rates remain high. The outcome has been that in most states there is a preponderance of young adults in the population, a situation that has placed strains on land resources, but also on employment markets, education and social organisation. A critical development issue throughout the Pacific is that of maintaining, let alone improving, present standards of living in the face of rapid population increases.

Migration

Migration has become a significant regulator of demographic change, particularly in smaller island states, where mobility is an intrinsic element of social, economic and political life. There are few areas of the world where migration has such an important bearing on the causes and consequences of population trends. Whereas, in the past, movement tended to be circular, often seasonal and usually over short distances, typified by male migration to plantations, permanent and relatively long-distance family migration has become more typical.

The depopulation of small islands and remote mountainous areas is widespread in the Pacific, though absolute population loss has only rarely been recorded (for example, Osborne 1966: 49). Employment opportunities and services (especially education) are concentrated in the urban centres and on small island states; where manpower and capital are often limited, centralisation is inevitable at some scale, hence selective rural–urban migration follows. The more educated have tended to migrate first and migrants have left many rural areas to take advantage of superior urban educational and employment opportunities. These two facets of migration emphasise and reinforce each other, hence migration results in the loss of the more energetic, skilled and innovative individuals, and this loss, which may reduce rural political bargaining power, business expertise and so on, is not compensated either by remittances or by some other form of trickledown effect from urban and national development. More recently, and primarily in parts of PNG (Connell 1997a), there are growing indications that migrants have moved away from inadequate rural opportunities, not because of the perceived superiority of urban opportunities, but out of increasing rural immiseration.

Migration is principally a response to real and perceived spatial inequalities in socio-economic opportunities. Though urban areas display high levels of open unemployment, migration continues because differentials between urban and rural average incomes remain substantial, thus migration hampers both economic growth and development. In most countries earning power is increasingly concentrated amongst urban bureaucracies while the absence of developed state mechanisms (such as progressive taxation, unemployment benefits and pension schemes), for affecting transfers of income, minimises redistribution towards rural areas other than through personal remittances. As early as 1980 the unpublished draft Solomon Islands Development Plan stressed these issues in detail, concluding that 'People in town have apparently become financially better off than rural people in recent years because they have been able to capture for themselves more of the benefits of monetary economic growth, largely at the expense of the rural people' (see Connell 1983e). Yet ultimately an economic rationale, real or latent, underlies most migration moves. In the Blacksands (Port Vila) and Ivane (Port Moresby) squatter settlements most migrants had arrived for employment and others for the superior education that might secure them a good job in the future (Vavine 1984; Mecartney 2001). Simply stated, in Port Vila, one of the most important reasons was '*long winem smal vatu from no gat rod long winim vatu long aelan* (to earn a little money since there's no way to earn money on the home island)' (Mitchell 2000: 172). For many that is reason enough. Growing inequalities, coupled with rising expectations, are the concomitants of increased migration.

A major influence on migration has been social change and growing expectations regarding what constitutes a satisfactory standard of living, a desirable occupation and a suitable mix of services and amenities. Aspirations involve imported goods (foods, clothes, vehicles, oil and so on), access to schools, hospitals, modern entertainment and travel, all of which demand some cash income. In parallel with changing aspirations and the increased need to earn cash, agricultural work has lost prestige and the relatively limited and declining participation of young people in the agricultural economy is ubiquitous. In Fiji, youths 'are taught to value white-collar occupations and detest farming' (Naidu 1981: 8). Since the Second World War, but especially in the post-independence era, the number of white-collar, and similarly prestigious, occupations has increased considerably, especially in the urban areas. The expansion of education has trained more people to be able to take up such jobs, and opportunities (however limited) are available to the young that had only exceptionally been available to their parents. Expectations are steadily rising but, at the same time, employment crises in many urban areas, growing rural populations, inflation, static (or even falling) commodity prices and the declining availability of land in some areas slowly increase the gap between expectation and reality. This increasing gap is one of the critical problems of development in the region.

Migration is strongly influenced by social, political and psychological variables. In Melanesia there is almost a ritualisation of migration; in places where initiation rites are dying out, migration is perceived as a modern form of initiation (Lewis 1980: 212). In a similar vein migration and exile are recurrent themes and symbols in much contemporary Pacific islands literature. Some notion of 'escape' is present in the movement of young men and women to avoid more or less arranged marriages, the social controls of a more traditional and highly visible rural life, or even as a respite from both stability and boredom.

Education is a major influence on migration. The generality of this education bias, despite the relatively short history of formal education in parts of the Pacific, suggests that it is likely to continue. Moreover, as access to the few new white-collar jobs becomes more difficult the 'diploma disease' worsens and the qualifications required for jobs increase. Attempts to develop a more appropriate curriculum have largely failed. In Kiribati, for example, vocational education in community high schools was strongly opposed by parents concerned that their children would be unable to obtain bureaucratic employment. As in Palau, educational systems create disdain for rural life because 'the exclusion of traditional skills and knowledge from westernized school curricula in many developing countries amounts to a constant tacit assumption that such things are not worth learning' (Johannes 1981: 148). In Blacksands some 7 per cent of families had moved there for better education for their children. Yet, though many had achieved it, some were forced to send their children back to the village for education as urban schools became full. A few families dismissed education, believing it no longer offered a guarantee of employment or the opportunity to improve living standards. Such families were usually financially disadvantaged and with little education (Mecartney 2001), hence the restricted access to education for their children was likely to perpetuate and accentuate the structure of uneven development.

Tertiary education is usually undertaken outside the home island, especially in the smaller states, further contributing to emigration. The Tongan anthropologist, Epeli Hau'ofa, has summarised the situation succinctly: 'Once you are educated, once your mind is expanded, subsistence on a remote little island is simply unacceptable. . . . Psychologically we are no longer islanders' (quoted by Dyson 1982: 120). Consequently,

> migrants seek in the West access to material goods, jobs in the industrial sector, better education for their young, and social mobility in a society they

> have believed free of the traditional barriers of rank and family status that made such mobility difficult at home.
>
> (Shore 1978: xiii)

Experiences and perceptions of the wider world, its values and its material rewards further underlie the migratory experience. Migration of the young to Koror has been summed up by one Palauan as the three Es: 'employment, entertainment and education' (Rehuher 1993: 21), a not uncommon situation.

Rather than migration being a result of individual volition, it is often 'linked', as village households organise the migration of family members to spread risks, diversify incomes or act as possible 'beach-heads' for further migration: at international level, something of a 'transnational corporation of kin' (Marcus 1981). The existence of kin in urban areas is a major influence and support. Not only do they provide demonstrations, or create images, of an impressive life style, they may also provide remittances (the visible monetary symbols of success), fares and accommodation for new migrants to the city. Many who initially visit towns for education or health reasons just stay on with relatives. Because of these kinds of explanations and justifications, and the *post facto* rationalisation of decision-making, migration is often best seen 'as an almost inevitable decision that they [villagers] will have to make sooner or later and once this view is accepted a sort of migration momentum develops' (Walsh 1982a: 7). This stimulates chain migration; in many contexts it appears a chain reaction.

Rising expectations in the wake of independence and growing materialism have exacerbated the shift from production to consumption, the decline of exchange and an increase in social tensions. The spreading taste for commodities has influenced work habits:

> [The] allure of modernity is strongly linked to city life. . . . The mystique, however, tends to be stronger than the rational expectations. It survives the disappointment of the latter. Whatever its frustrations and degradations, the city continues to be the place where things are happening, where there is movement and a sense of future.
>
> (Berger *et al.* 1974: 128–9)

For many in the Pacific, the largest cities and the metropolitan countries exercise the same allure, offer the same sense of future and simply validate migration.

Contemporary migration from the Island Pacific

International migration to the metropolitan nations on the fringes of the Pacific is primarily a Polynesian phenomenon. Many people from Tonga and Samoa have moved either to New Zealand (whence some have gone on to Australia) or, increasingly, as the New Zealand economy has stagnated and immigration restrictions have become tighter, to the United States. In the larger countries of Melanesia emigration is conspicuous by its absence, though there has been significant emigration from Fiji, especially of Indo-Fijians. This dramatically accelerated after the 1987 and 2000 coups, with migration to Australia, New Zealand and also Canada. The former US territories of Micronesia – Palau, Marshall Islands and particularly FSM – have increasingly exhibited similar trends, accelerating in the latter two cases since the signing of Compacts of Free Association with the United States. For even longer Kiribati and Tuvalu have been characterised by migration, but of contract labour – mainly to Nauru (for employment in the phosphate mine) or to work in the international shipping industry (for which both countries have training schools) – hence return migration is normal and the impact on national population change less significant.

The scale of international migration currently appears to be more closely determined by the vicissitudes of the international economy (and specifically that in the United States and New Zealand) rather than of the domestic island economies, thus the scale of future international migration is unpredictable. However, it is growing fastest in Micronesia, where there are no legal restrictions to emigration (at least to the United States and its territories, notably Guam and the Northern Marianas), despite limited language ability and skills making it inevitable that most migrants can only obtain low-paid 'entry-level' employment.

Tonga and Samoa share a history of substantial emigration lasting more than three decades; in each of the three Polynesian states population growth and urbanisation have been slowed by emigration, not least because of the loss of the most fertile groups. Migration (and the extent of return migration) will remain the most important influence on urban expansion (or contraction) in the immediate future. Though both Tonga and Samoa experienced some emigration in the early post-war years, large-scale movements effectively began in the late 1960s. A combination of reasons, including recognition of substantial income differentials between Polynesia and such metropolitan states as New Zealand and the USA, increased expectations of superior education and health services, improved transport links (especially air transport), a relative reduction in transport costs and growing population pressure on domestic resources, all stimulated mobility (Connell 1983f, 1983h). In the early 1980s migration slowed because of adverse economic conditions in New Zealand and migration streams increasingly shifted towards America, often via American Samoa (Ahlburg and Levin 1990). Current estimates suggest that almost as many Samoans and Tongans live overseas as at home, in countries where their future is intricately related to economic prosperity, recession, restructuring and the nature of the employment market. At no time during the past quarter of a century has there been substantial return migration; moreover, in Samoa prospects for emigration were so poor at the start of the 1980s that the 'broken dreams' of potential migrants contributed to a significant rise in youth suicide (Macpherson and Macpherson 1987). At a national level the economic future of the two states partly hinges on the continued flow of remittances, and hence on some continuity of migration (Ahlburg 1991; Connell and Brown 1995). Where there has been return migration to Tonga and Samoa, much of it appears to have been to Nuku'alofa and Apia rather than to the rural areas that were generally the migrants' original home areas. The possibility of blocked migration in the future, a situation ever present in public debate (Macpherson 1992; Shankman 1993), emphasises the potential problem of a high rate of natural increase in case emigration is substantially reduced, since there is now a 'culture of migration': emigration is normal, expected and anticipated, and an important element in national social and economic systems.

Tuvalu and Kiribati are rather different, despite labour migration for more than a century. Though Tuvalu has experienced relatively little permanent emigration, in 1991 there were more than 1,200 Tuvaluans overseas, over 10 per cent of the national population. The largest number of these (735) were on Nauru where the phosphate workings were reaching the end of their life, ultimately necessitating the return of these contract workers to Tuvalu. In Kiribati, where there were some 3,000 I-Kiribati overseas in the mid-1990s, the situation is almost identical to that in Tuvalu. Kiribati and Tuvalu are two of the very few countries in the world where the local population is trained for work overseas, but the future of overseas migrants – as seamen, miners, students or contract workers in New Zealand – is insecure. Both countries face the more inevitable and perhaps more imminent return of a very significant proportion of their population. As in the larger Polynesian countries, return is likely to place particular pressure on the urban centres where population densities and their impact on very limited resources are already considerable.

The signing of the Compacts of Free Association, giving Micronesians unrestricted access to the USA, quickly encouraged movement by some of those with skills who could not find government employment at home, led students to seek permanent residence in the USA, and stimulated rapid migration by the Chuukese, in particular, to Guam and Saipan. A more comprehensive pattern of international movement accelerated and spread in the 1990s (Hezel and Levin 1996). Overall more than 10,000 FSM citizens may now be outside the country. A broadly similar situation is also true of Palau, though it has been much longer established, and of the Marshall Islands (Ogden 1994: 256–60; Smith 1996) where emigration was particularly rapid in the 1990s. The migration process in Micronesia is becoming increasingly similar to that in other parts of the South Pacific: a steady outflow, growth of relatively permanent urban communities overseas (beyond student groups) and the compensatory flow of remittances. As elsewhere a culture of migration has gradually become established.

In rather different ways the national and urban futures of each of the Polynesian and Micronesian states are bound up with the present and future of international migration and the ability of these countries to seek out new potential destinations overseas. All have retained interests in overseas employment opportunities. Both Tuvalu and Kiribati have experienced resettlement overseas, Tonga has leased agricultural land in Asia and in the 1980s the Jackson Committee on Australian foreign aid recommended that special provision be made for migration from Tuvalu and Kiribati because of the improbability of conventional aid contributing to sustainable national development. More recently Tuvalu's Prime Minister stressed that his country was continuing to seek employment and migration opportunities in Australia and that Tuvalu 'would not take no for an answer' on the provision of either employment or education opportunities, regarded as necessary for economic survival and potentially critical in the face of future sea-level rise (Connell 1999). There has already been a brain and skill drain (and a 'brawn drain' of sportspersons) from each state and all are anticipating substantial future emigration. Migration has reduced the level of open and disguised unemployment, although it has also contributed to a loss of skilled human resources in the 'modern' sector. Rising material consumption levels following migration have generated increased demand for consumer goods, a demand that can only be satisfied by further migration, as migrants bypass the small towns of the Pacific to seek superior living conditions overseas.

The urban impact

Evaluation of the urban impact of migration in the Pacific islands region is not easy, both because it is difficult to distinguish from that of other socio-economic changes and because a growing proportion of the urban population has been born in urban areas, even in Melanesia. The most obvious category of permanent urban residents are the inhabitants of villages now incorporated into urban areas, such as Hanuabada in Port Moresby, or the descendants of labour migrants from the early colonial era, such as Niueans in Apia (Samasoni 1987). The original relationship to the land established by the colonial authorities over one hundred years ago remains an important factor influencing the very different degree of commitment to permanent urban residence. Thus the Indo-Fijians of Suva, who are more committed to urban residence than Fijians (Nair 1980), exhibit some similarities with the poorest migrants in Port Moresby, who may have limited opportunities for access to land at 'home', yet neither of these two groups can genuinely be considered a proletariat. Migrants and their children from some of the poorer rural areas, such as the seemingly permanent urban residents originating from parts of the East Sepik Province and from the Gulf Province in PNG (Morauta and Ryan 1982), are firmly established in town. Rather differently, so too are bureaucrats and other elite workers who have skills that are inappropriate in rural areas and economic and social expectations that cannot be met there. A growing differentiation has emerged between those permanent urban residents who are relatively poor (including some urban villagers and the migrants from poor rural areas) and others who are well off. This is most apparent in Melanesia and, to a lesser extent, Micronesia.

Until no more than a quarter of a century ago, factors that discouraged urban residence were more important than those encouraging permanence, and they remain of some significance. No social security or welfare systems provide for

Plate 3.1 Hanuabada, Port Moresby, Papua New Guinea. © John Connell

urban living when an individual is unemployed or has retired, whilst, in Melanesia especially, some urban workers are in 'tied' accommodation, associated with the job. Many live in squatter settlements where their tenure may not be legal. Urban incomes are not always adequate to support a family, a factor which has also discouraged migration, even in PNG, where urban wages have been high relative to many other developing countries, with a high urban minimum wage set in colonial times. Hence there has been a declining rate of urban employment creation, a situation that remained unchanged until 1992 when a new government decided to reduce the urban minimum wage. Whilst the policy was intended to encourage industrialisation, and stimulate other components of the private sector, it increased urban social problems, discouraged rural–urban migration and brought PNG closer to the situation elsewhere in Oceania where urban–rural wage differentials are less substantial and the wage benefits of migration less evident (Levantis 1997a). Finally, in PNG especially, the heterogeneity of urban life (incorporating migrants from a variety of traditionally opposed language groups and regions) contributes to some degree of insecurity, especially where unemployment, violence and crime also occur. Such problems are fewer in the other island states.

Other characteristics, including the decline in the rate of creation of new formal sector urban job opportunities and a reduced turnover in the workforce, have tended to encourage a more stable urban population. Urban conditions have improved substantially for the employed as urban wages have increased; urban unions and militant urban workers in several countries have secured superior urban wages relative to the incomes possible in rural areas (especially where there have been static commodity prices). This is particularly true of public service employment, where there has been relative job security, high wages, considerable freedom, pleasant working conditions and access to 'fringe benefits', such as telephones and vehicles. Average wages and salaries in the public sector are everywhere greater than in the private sector, sometimes as in the former TTPI states considerably greater, alongside superior superannuation and retirement payments. (Such advantages, most obvious in the capital cities, have ensured that the public service is the most sought after area of employment, second only to being a Member of Parliament.) In a number of rural areas cash-earning opportunities have declined and transport links worsened, particularly in the smaller islands and even parts of some larger islands, such as northern Malaita (from where migrants in Honiara, even before independence, commonly used the expression '*Mi lusim passport bilong mi*' (I have lost my passport) as an expression of length of urban residence and the problems attached to return).

Migration has resulted in intermarriage across ethnic and linguistic groupings; such couples often have fewer opportunities in rural areas (depending on the structure of land tenure), experience some social problems in either of their home rural areas (as loyalties and interests become divergent and conflicting) whilst they have children with very few ties (particularly linguistic or emotional ties) to the rural areas (Connell 1988b). A quarter of a century ago, a quarter of the children born in the large towns of Port Moresby, Lae and Rabaul had never visited their 'home villages' (Garnaut *et al.* 1977) but primarily because they were too young to have had an opportunity to go 'home'. None the less, over time, urban permanency has become more important and for some is now seen to be irrevocable (Ryan 1989). Yet 'their very permanence disadvantages them; high dependence ratios, little or no support from the rural economy and no opportunity to move away from town when poverty, rising unemployment, old age or social disorder make life in town difficult or unpleasant' (Morauta 1980: 297), at least for those who were, in one way or another, 'trapped' in town. In Blacksands, insecurity over land tenure and employment ensured that migrants contemplated return, but most realistically believed they would remain in town for their children's sake (Mecartney 2001: 80). The combination of growing urban permanency, high unemployment and increased expectations have put considerable pressure on urban services.

Increased female and family migration has contributed to balancing sex ratios in most Melanesian towns where, until recently, there were considerable gender imbalances. In Port Moresby the masculinity ratio (males per hundred females) fell from 185 to 139 between 1966 and 1977; it was 125 in 1990. In the mining town of Tabubil it was 150. Similar changes were recorded much earlier in Fijian urban areas where the sex ratio has long been evenly balanced (Parliament of Fiji 1979: 17–18). In Polynesia and Micronesia there has never been a significant gender imbalance. The movement towards more balanced sex ratios has resulted in natural increase becoming a more important determinant of urban population increases. Although ambivalent attitudes to urban life and residence persist, migrants have stayed longer in town, women have come in increasing numbers, thus producing and raising families in town. Migration from rural areas is more likely to be permanent; this permanence is in urban areas in the Pacific (and there are now third and fourth generation urban residents) or in metropolitan nations.

Against the city

The benefits of migration to urban areas were particularly apparent in the early post-war decades when migration and urbanisation were assumed to be concomitants of economic growth and development. Migration enabled individuals to obtain higher education and technical training (or merely gain experience in a different milieu) and contribute to urban and national development through their labour. Many new skills were also of value elsewhere, in the event of return migration. Not only did migration have positive benefits in the towns but it contributed both to a reduction in population pressure on resources in rural areas and to natural resource development (usually in mines and plantations) elsewhere. Consequently, until relatively recently, urbanisation in the Pacific was viewed positively. In PNG, where urbanisation was relatively late, towns were seen on the eve of independence as centres of national social, economic and political development (Chapter 2). While there were dissenting views, there were few urban problems and no real hostility to urbanisation in the region.

Growing concern over increasing urbanisation and its apparent permanence was a function of a more visible informal sector, rising urban unemployment, the growth of shanty towns, urban crime and other social problems, the kinds of issues that brought Michael Somare's queries over the nature of urbanisation in PNG (Chapter 1). As cities continued to grow and new attention was focused on urban health and environmental issues, perceptions of increasing urban social disintegration (necessitating intermittent curfews in the worst Papua New Guinean instances) deterred favourable perceptions of urban life. Once underlying that concern was an ideological conviction that the Pacific way of life was a rural one, and should continue to be so, and that urbanisation was not part of the Pacific way. Such ideologies have largely disappeared, though they rhetorically surface in the words of urban politicians and others (Koczberski *et al.* 2001), matched by attempts to reduce the extent of urbanisation.

In some Pacific countries (most obviously Fiji and Solomon Islands) national policy has briefly been orientated towards the development of growth centres away from the capital city – both to take advantage of local resources and to spread urbanisation and economic development more evenly (Sevele 1979: 40–1; Stavenuiter 1983). The experience of such growth poles and of more general policies of decentralisation has not been encouraging. The high costs of direct government expenditure on establishment and subsidies have often been prohibitive, skilled workers have been reluctant to leave urban areas and development has tended to be capital intensive. Ultimately however there have been relatively few such schemes, since there has been little enough economic growth to decentralise; bureaucratic inertia and the shift towards deregulation have killed off the few that existed. Only where there has been new natural resource exploitation, primarily at the new mining centres in PNG, have new towns emerged in the post-independence years.

The more rapid urban growth that often coincided with independence stimulated new forms of national development, especially in Port Moresby, where impressive structures – such as, belatedly, a new National Parliament, a museum and so on – marked the political transition. In PNG there has been some discussion of establishing a new national capital, more central to the needs of an independent state, but only in the FSM has a new capital been completed. The small national complex at Palikir, outside Kolonia, the tiny capital of Pohnpei state, comprises no more than the parliament and some offices. In Palau plans were approved and construction underway on a national capital on Babeldaob but activity was held up by the collapse in 1996 of the KB Bridge linking Koror with the 'big island'. Almost all countries have taken some interest in decentralisation and resettlement. In the atoll states of the Marshall Islands and Kiribati there has long been interest in decentralisation, with both countries having government departments with that as a key objective.

Kiribati especially has achieved some degree of success. For more than 60 years attempts were made to decentralise population from the Gilbert Islands group to the Phoenix Islands and Line Islands. Recognition of population pressure in Kiribati in the 1930s prompted the colonial administration to consider resettlement as a solution. The first attempt was the Phoenix Islands resettlement scheme of 1937, followed by post-war schemes, initially to resettle Banabans to Fiji and, subsequently, the Phoenix Island settlers and others to the Solomon Islands (then also a British colony). Finally the relatively thinly populated but remote and isolated northern Line Islands were identified as the most important internal destination for an overflowing population, since the three islands of Tabuaeran (Washington), Teraina (Fanning) and Kiritimati (Christmas) have nearly 60 per cent of the nation's land resources. Between 1985 and 1995 the Line Islands' population doubled to almost 6,000. In 1996 the focus shifted again to the Phoenix Islands as the capacity of the Line Islands to absorb more settlers was at an end, though the Phoenix Islands possess enough natural resources for just a thousand migrants. Despite a very deliberate attempt to achieve population balance through resettlement, it has been an expensive and inefficient solution to population pressures elsewhere.

No other countries have groups of thinly populated islands (especially where land tenure is not an issue) that provide potential targets for resettlement, hence decentralisation policies have merely sought to encourage return migration to 'home' islands and regions. In the Marshall Islands, the severe problems that followed the urbanisation of Ebeye atoll in the 1960s (Chapter 2) prompted several direct attempts to persuade people to return to their home atolls; eventually 'Operation Exodus' was established to reduce numbers from nearly 5,000 to between 3,000 and 3,500, encourage family planning and increase per capita incomes by reducing population and maintaining full employment

(Vitarelli 1967). The population of Ebeye fell, but only temporarily (Connell 1983c: 25–6). In 1975 a new exodus programme saw a few hundred of the unemployed return to outer islands voluntarily after 'persuasion' by leaders (*iroij*) had taken place. Forced removal is not an option in a country where freedom of movement is a major political issue.

The situation in the Marshall Islands was thus similar to that elsewhere in Oceania, where policies were more obviously in opposition to urbanisation than in support of decentralisation and resettlement. In states like Kiribati and PNG, the colonial control of population movement continued until post-colonial times. In Kiribati early administrative regulations were designed to slow the rate of urban growth and minimise the various problems associated with over-rapid urbanisation. Entry to Betio, the most urbanised island in South Tarawa, was restricted in theory to only those having a legitimate reason for being there and magisterial permission was required to board vessels travelling there. Although such administrative controls constituted the only direct attempt to manage the 'disastrous shift of population' (Fox 1976: 57), some officials saw rural development as the only real solution to urban problems. As early as 1969, the Gilbert and Ellice Islands development planner Tony Hughes argued that:

> So called remedies for urbanization problems suffer from confusion about the nature of the problem. The result is that the symptoms are treated rather than the cause. Urbanization is a problem only because people 'urbanizing' make impossible demands on their own human resources and on the physical resources of the town. The allocation of more . . . usually only makes things worse by accelerating the inflow of people to make use of them. Politically loaded choices are made in resource allocation to keep the urban voters quiet.
>
> (Cited in Bailey 1984)

But the administration continued to centralise programmes and institutions in South Tarawa, and even subsidised urban amenities through much higher per capita expenditure in the town than elsewhere (Fox 1976: 57–8). Early attempts to define and devise a migration policy thus met with little success. Subsequently, continued urbanisation resulted in the government of the independent state going beyond resettlement policies to discourage urban migration more actively, and design a specific public sector employment policy intended to share out work 'in a more equitable manner and, at the same time, persuading people to either go home to their own islands or stay there in the first place' (Government of Kiribati 1981: 2). Policies designed to allocate existing employment more equitably included making all unskilled jobs within government and statutory bodies on South Tarawa three-year contract jobs only; recruitment was from the outer islands (or from 'true Tarawa' people) on a quota basis and at the end of the contract workers were expected to return home (Connell 1983b). Eventually this posed bureaucratic problems, beyond the gains that were achieved in terms of equity and decentralisation, and in the mid-1980s the policy lapsed.

In PNG, above all, even interest in achieving a different spatial distribution of the population, through regional and rural development, was only infrequently present in national development plans and policies, though there has sometimes been strong opposition to urbanisation and especially squatter settlements, through prejudice rather than any notion of rural development policy. Squatter settlements are widely perceived to be haunts of the feckless unemployed (though there is little evidence of this being particularly prevalent); a significant proportion of employment is in the informal sector and this is rarely regarded as appropriate or genuine employment (Chapter 4). A correlation between urban crime levels and recent migration is frequently voiced but also lacks demonstrable proof; there is no evidence that crime levels or gang membership are more obviously correlated with settlement residence (Goddard 1992, 2001; Nibbrig 1992). It is not implausible however that social disorganisation and crime are a function of substantial inequalities in access to land, housing and other services in the largest cities. Certainly some PNG criminals themselves have pointed to crime being a response to both inequalities and their perceptions of corruption and other 'white-collar' offences that have gone unpunished. Crime tends to exacerbate the deprivation experienced in settlements, and to make them a source of greater concern amongst other urban residents.

Settlements are consequently also perceived to be inappropriate phenomena in capital cities and disfigurements of urban landscapes (Koczberski *et al.* 2001). Even before PNG independence, as migration to towns increased in volume, there were suggestions that a pass system be introduced; as early as 1971 the Morobe District Councils Conference wrote to the United Nations requesting that the UN Universal Declaration of Human Rights be changed to enable passes to be introduced. Whilst nothing came of this or similar early proposals, by the end of 1977 opposition to migrants in the North Solomons (Bougainville) Province had intensified to the extent that at least three groups of squatters were repatriated by plane to the New Guinea mainland (Talyaga and Olela 1978), despite the fact that the national constitution provided for freedom of movement. The North Solomons Provincial Government (along with Morobe and East New Britain Provinces) further sought to introduce compulsory identity cards and restrict those seeking work to a maximum of 30 days in the province. As Denoon (1980: 280) observed: 'The attraction of closed towns is very obvious. . . . With one bound we rid ourselves of poverty, ignorance and disease; modern towns shed their nasty reminders of underdevelopment. . . . Closed towns are the millennial dream of the property-owning, rate-paying middle class' (see also Connell 1985b: 107–15). Dreams never became realities.

Elsewhere the problems of urbanisation have usually been less evident than in Port Moresby and the response has been less dramatic. The only direct opposition, until very recent years, to squatter settlements in the Solomon Islands was in 1981 when the Cabinet decided to remove the long-established Fishing Village (Vaivila) settlement from the foreshore on the grounds that it was a health problem, overcrowded, unsightly (in a capital city) and that there were security problems, though it was never relocated (Connell 1983e). The only Honiara Town Council Development Plan sought to reduce 'immigration' (although potential mechanisms for achieving this objective were unstated), supporting national government policies 'providing for a more even distribution of employment opportunities throughout the country' (though other than vocal support it is unclear how this would occur), a greater use of planning regulations, encouraging family planning and publicising 'the desire of the Town Council to limit population growth' (Honiara Town Council 1988: 127). However, no particular policies were implemented to achieve these objectives. Such typically vague and occasionally worthy objectives, associated with reducing rural–urban migration and improving urban management, here as elsewhere were indicative of the limited resources available to urban authorities for managing the process of development.

In PNG opposition to urbanisation has continued, from both urban authorities and influential leaders. There is, however, as there has always been, a paradox:

> Everyone blames the squatters for the sudden increase in crime. Some have petitioned the government to get rid of these settlements, yet politicians visit these same places at election time with money, beer and rice bags to win their vote.
>
> (*Times of Papua New Guinea* (hereafter *TPNG*), 25 October 1984)

At different times most of PNG's urban authorities have sought to eject at least some squatters and/or the unemployed. In 1984 Madang tried to eject squatters because of problems of law and order and land shortages (using the argument that settlements were on land required for gardens, urban expansion and resettlement of the local Madang population). At the same time a committee of inquiry recommended 'that the provincial government carry out a review of all government land with a view to creating an integrated land development scheme and making provision for controlled settlements for low income earners and unemployed people' (*TPNG*, 5 July 1984). Whilst it proved impossible to eject squatters, the more positive recommendations on integrated land management were equally unsuccessful.

More recent attempts elsewhere have produced similar results. In mid-1991 the Morobe premier sought to eject all illegal settlers from Lae:

> My plan to eject settlers is a genuine one for the sake of my people who want a trouble free environment for their children. . . . As for Papua New Guinea which has a large area of land, I do not see why people should move from province to province. I am sure there are better things to do in their own villages or towns.
>
> (*TPNG*, 6 June 1991)

Not surprisingly this was not the settlers' perception. At the same time in Rabaul illegal settlers were to be evicted for the same kinds of reasons. The East

New Britain premier paternalistically stressed that the exercise was being undertaken for the good of both the province and the settlers:

> We want to help the settlers so that they can have dignity and that they can grow their own cash crops in their respective villages. Villages are still the best and least expensive places to live. Furthermore we want settlers to understand that they were not born to be slaves in towns. However they are making themselves slaves.
>
> (*TPNG*, 29 August 1991)

Similar policies there, and in other PNG towns, especially the largest ones, have continued throughout the decade (Connell 1997a; Goddard 2001; Koczberski *et al.* 2001) in the guise of achieving order and cleanliness, reducing crime and unemployment, freeing land for business development, and demonstrating that the state was not weak. Violence and destruction – against settlements – have become instruments of urban planning as urban problems become increasingly complex and unmanageable. The former Governor-General, Sir John Guise, well exemplified concerns on the growth of settlements over a decade ago:

> A state of emergency for the National Capital District is long overdue. It is in the city of Port Moresby where ordinary people live in constant fear of their lives and property. We cannot and will not contain the criminal situation in this land unless the government stops the urban drift of people from the villages. The government needs to be courageous in weeding out the unemployed people in Port Moresby as well as in other urban centres back to their home provinces and villages.
>
> (*TPNG*, 3 January 1991)

The list of similar statements, attempts at eviction of squatters and opposition to rapid urbanisation in PNG is a long one. In Vanuatu as well, settlements in official discourses and newspapers have also been seen as places of 'squatters' and sources of criminals, 'a blight in paradise and a place of transgression', from where people should return home (Mitchell 2000: 192) but no direct action has been taken. Though there has been some justification in terms of a need for rural development, the principal opposition to settlements has come from established residents, fearful of crime and loss of urban amenity and occasionally concerned over land shortages. It has only exceptionally been possible to translate opposition to settlements into removal of settlers (in part because of both constitutional problems and the legality of settlers). Settlements, unemployment and crime are discursively linked by politicians, police spokespeople and 'monotonously repeated in the daily press', or on radio stations, alongside routine contextualisations about the inevitable consequences of urban migration and poverty (Goddard 2001: 4) that invariably lack substance. At the very least there is a considerable diversity of settlements and settlers, though the claims of settlers for financial support, through service schemes and social amenity provision, are unlikely to be met – except at election times – and are usually routinely rejected. Such intermittent ambivalence, and weak governance, has meant that, even though there has been an enormous amount of debate, it has not been possible anywhere to formulate an effective policy to discourage, remove (or even upgrade) squatter settlements.

By contrast to the widespread opposition to urbanisation – at least in the form of settlements – there has been minimal support for the rights of settlement residents, other than from themselves. In 1990 the Papua New Guinea Attorney-General, Bernard Narokobi, suggested that the government consider some social security scheme as a possible solution to curbing the growing urban crime problem: 'We should provide some sort of social security benefits to the unemployed who have sincerely tried but cannot find employment. The dole would be restricted to second and third generation citizens born in major towns and cities' (*Post Courier* (hereafter *PC*), 4 December 1990). Though Narokobi had restricted this possibility to long-established residents, he was strongly criticised for a policy that might encourage more migration rather than for focusing on the necessity to encourage rural development. However, such forays into the considerable complexity of urban welfare provision have been rare. The pervasive opposition to urbanisation has delayed and discouraged the development of coordinated plans for urban management.

Opposition to urbanisation has focused primarily on its symptoms – the expansion of squatter settlements – and rarely (obviously, in part, because it is much more difficult) on the rationale for rural–urban migration and urban growth. Whilst the effects of urbanisation are often seen as social problems, their causes are economic, a result of emerging inequalities between urban and rural life, and within the towns. Authoritarian or coercive controls over migration address the symptoms and not the causes of resource misallocation and regional disparities; 'urban problems may be the manifestation of rural underdevelopment' so that 'closure of the towns will merely complicate the problem by obscuring it' (Denoon 1980: 280) and further emphasise rural poverty and rural–urban inequalities.

The extent to which disinvestment in urban services, as opposed to straightforward opposition to settlements, is a disincentive to rural–urban migration is nowhere clear, least of all in Melanesia. The most acceptable conclusion appears to be that, in PNG at least, 'the government's control of urban migration through budget-cuts on urban services to date does not seem to have any effect on the continuing migration into the urban areas' (Aruga 1992: 13). Mismanagement may have been a subsequent incentive. A more probable 'deterrent' to rural–urban migration would be rural development. While urban wages remain high, industrialisation is encouraged, some rural areas are distant from market access and services are concentrated in cities, even successful rural development policies can only slow rather than reverse the extent of rural–urban migration.

The institutional response (or lack of response) to urbanisation has been broadly similar, though more limited, than in other developing areas. Despite official concern over managing urbanisation, alongside an expressed focus on rural development, there have been few real attempts to constrain urban development. Urbanisation has been limited, primarily because of the existence of reasonable income-earning opportunities in many rural areas (or overseas) rather than through rural development policy and practice. Despite some policy orientation towards rural development, expenditure in urban areas is quite disproportionate to urban population sizes. What is perhaps surprising is that policy formation directed to mitigate urban bias and too rapid urban growth has not been possible in conditions where most politicians have extremely strong rural economic ties, where an alienated urban elite is barely apparent, and where migrants are at least as conservative and poorly organised as elsewhere in the developing world (and hence place limited pressure on urban service provision).

Migration continues to be perceived as a means of escaping limited local economic opportunities and of diversifying or transferring human resources into more productive areas. Although virtually the only jobs for uneducated, unskilled migrants are poorly paid, demand has been substantial, especially from remote areas where income-earning opportunities are very limited (Boyd 1990). The educational filter that has restricted access to scarce skilled and well-paid jobs has served to differentiate migration streams and stratify the social and economic structure of the cities. Yet migration has also increased options, diversified earning opportunities, enabled the purchase of a range of new goods, improved access to services and exceptionally led to successful careers. The shift of migration from circularity towards permanence, as job opportunities decline, rural opportunities worsen and population grows, has worsened rural–urban inequalities, themselves often emphasised by 'urban bias' in government policy, and also worsened intra-urban inequalities. Slowly, and not very surely, population growth and migration have contributed to a more urban future for the Pacific.

4 Economic development

Towards the city?

> Most Samoans who work in the modern sector either at home or in New Zealand, are regarded by their relatives, if not themselves, as bridges through which *fa'a Samoa* [the Samoan way] is extended to tap new resources. Thus kinship pressure on Samoans living and working in urban settings continues to be dictated to by the concepts of *fa'a Samoa*. Great ideological issues are related to economic transformation.
>
> (Meleisea 1992: 63–4)

As in other small economies, development in island states is constrained by the limits of small size. Well-known influences include: remoteness and isolation (resulting in high transport costs to markets); diseconomies of scale (because of small domestic markets); limited natural resources and a narrow production base, substantial trade deficits (because of dependency on metropolitan states); few local skills; vulnerability to external shocks and natural disasters; and a disproportionately high expenditure on administration and dependence on external institutions (such as banks and universities) for some key services. Moreover, political systems have sometimes been fragile, ecological structures are vulnerable and economies lack diversity. Small islands, a paucity of natural resources and remoteness, even within countries, have hampered the ability to compete in the global economy. Consequently island economies have traditionally specialised in a very narrow range of agricultural exports, such as copra and coffee. Export diversity is less than in any other world region. Set against this range of disadvantages, the comparative advantages of smallness and isolation are few, other than the retention of a degree of cultural integrity (Connell 1988b). Although the region does not suffer the absolute depths of poverty experienced in other parts of the developing world, it does have serious social and economic problems. Economic growth has been disappointing since independence, expanding populations have intensified pressure on lands and seas and political instability has increased.

The regional economy

The Pacific is subject to recurrent natural disasters, such as cyclones, volcanic eruptions and earthquakes. Separate cyclones in 1987 were estimated to have caused damage of the order of 150 per cent of GDP in both Vanuatu and the Cook Islands. Cyclone Kina caused widespread damage, flooding and deaths in Fiji in 1993. Some 5,500 became homeless, transport systems were disrupted, crops and farm equipment destroyed and total damage was estimated at F$170 million (US$130 million). A year later, Rabaul, then the fifth largest town in PNG and built within a volcanic caldera, was obliterated by the eruption of two volcanoes; though only four lives were lost, it was officially declared the nation's worst ever disaster. A state of emergency lasted for six months, more than 90,000 people were displaced and six years later reconstruction plans had yet to commence. Drought, a consequence of El Niño, devastated PNG in 1997, closing the two large mines at Porgera and Ok Tedi (since rivers were no longer running), reducing the production of export crops such as coffee, and resulting in considerable starvation and death. Pacific economies are all vulnerable to external shocks, either natural or through the price fluctuations of commodities. An additional 'external shock' has been the closer association with metropolitan states which has encouraged expectations of a standard of living beyond the likely means of most of the islands.

Most countries, other than Melanesian states, have benefited from aid flows and remittances, enabling them to run big current account deficits, maintain substantial bureaucracies and undertake relatively large public sector investment programmes of a kind that could not otherwise be financed. The Pacific islands are the most heavily aid-assisted region of the world on a per capita basis. The public sector dominates formal economic activity everywhere and, despite the availability of external resources and sound financial management, national incomes have largely stagnated since the 1980s. The combination of high rates of out-migration and large aid and remittance flows, with the resultant growth of a bureaucracy concentrated in the towns, has resulted in the smaller states being conceptualised as MIRAB (migration, remittance, aid and bureaucracy) economies (Bertram and Watters 1985; Connell 1988b) in that they are heavily dependent upon external support rather than on domestic production and trade. Though this designation is unwelcome within the Pacific, because of its suggestion of mendicancy, it is not inappropriate despite questions over its sustainability (Sato 1997; Poirine 1998), especially in Polynesia.

Fiji and the three other Melanesian states are relatively rich in resources, have achieved a considerable degree of economic diversification, and may not be permanently dependent on aid (see Table 4.1). PNG, particularly, has substantial minerals, timber and fish. Each of the Melanesian states has achieved some degree of economic diversification, though high population growth rates, land tenure disputes, political instability and a very low capital base make an early reduction in aid levels unlikely. The prominence of PNG and Fiji in the Pacific is apparent from their populations and land areas, but also in their having

Table 4.1 Pacific island economic characteristics

Country	*Area (km²)*	*EEZ (million km²)*	*2000 Pop'n estimate (1,000)*	*1988–98 Average GDP growth (%)*	*Principal domestic exports*
FSM	701	2.50	118.8	–4.0	Copra, fish
Fiji	18,300	1.30	824.9	2.4	Sugar, garments, coconut oil, gold, ginger, fish, timber
Kiribati	810	3.60	90.7	2.4	Copra, fish, seaweed
Marshall Islands	181	1.94	51.8	–0.9	Coconut oil, fish, trochus shells
Palau	490	—	19.1	—	Fish
PNG	462,243	3.12	4,790.8	–0.6	Oil, gold, copper, timber, coffee, palm-oil, cocoa, copra
Solomon Islands	28,370	1.50	447.9	3.9	Timber, fish, palm-oil, copra, cocoa
Tonga	650	0.70	100.2	2.3	Squash, vanilla
Tuvalu	26	0.90	9.9	n.a.	Stamps, copra
Vanuatu	12,190	0.78	199.8	1.6	Copra, beef, cocoa, timber, shells
Samoa	2,934	0.13	169.2	–0.1	Taro, coconut cream, wire

Source: Forum Secretariat (1996: 3)

85 per cent of the region's GDP, over 90 per cent of exports and 80 per cent of imports. They also house the three island Pacific universities and most key Pacific regional organisations. The two larger Polynesian states of Tonga and Samoa provide an adequate level of subsistence, though this does not meet the current expectations of some citizens, resulting in considerable emigration. Tourism has grown only slightly and they have small Exclusive Economic Zones (EEZs). Both countries appear likely to have a long-term dependence on foreign aid to 'top-up' their economic performance. The much smaller and resource-poor countries of Kiribati and Tuvalu require aid just to maintain basic services, have little conventional development potential and are unlikely to escape a future of long-term dependence on the wider world. The final group are the three former American entities of FSM, Marshall Islands and Palau for which aid constitutes the bulk of their national incomes, despite attempts to reduce aid dependence and stimulate the private sector, and where governments dominate the economy.

Subsistence agriculture and fisheries play a major role in the employment structure of most independent states (Table 4.2), especially in Melanesia where PNG, Vanuatu and Solomon Islands all record levels above 70 per cent. In

Table 4.2 Employment by sector

Country	*% of labour force*			*Government workers*		*Ratio of central govt wages to mean per cap. GDP*	*Unemployment rate*
	In agriculture	*In industry*	*In services*	*As share of non-agric. employment*	*Per 100 inhabitants*		
FSM	48.0	13.0	38.0	n.a.	6.3	3.2–9.1[a]	16.2
Fiji	44.1	13.8	42.1	49	6.0	4.2	7.5
Kiribati	73.8	6.9	19.3	35	4.7	10.9	—
Marshall Islands	21.5	20.9	57.6	25	6.9	4.0	12.5
Palau	14.6	23.9	61.6	n.a.	n.a.	n.a.	7.0
PNG	79.9	5.3	14.8	36	2.0	6.1	11.9
Samoa	63.5	5.5	31.0	42	2.4	4.6	2.1
Solomon Islands	82.0	9.0	9.0	43	4.0	3.8	—
Tonga	49.1	12.2	38.7	48	5.1	2.6	9.0
Tuvalu	41.1	3.3	55.5	n.a.	8.0	4.3	—
Vanuatu	74.4	3.4	22.2	32	3.0	5.2	—

Sources: Forum Secretariat (1996: 6); UNDP (1999)

Note
a Range of national and state figures, 1994.

Micronesia, only Kiribati with 74 per cent of the workforce engaged in subsistence activities, might be regarded as 'typical' of the Pacific region. The remainder of countries in the northern Pacific vary from the regional pattern, having economies subsidised by the United States, hence budgetary transfers subsidise a high level of employment in the public sector. Only in Fiji is there industrial development of any significance. Such limited diversity emphasises the massive importance of agriculture and fisheries throughout the region, and thus the rural nature of most economies. More than half the Pacific states were facing some form of economic crisis by the mid-1990s. All continue to experience a range of persistent difficulties in trying to overcome the diverse barriers to sustainable economic development. These economic experiences and the differences between national economies are of considerable significance for the structure and extent of urbanisation.

Primary production

No images of the Pacific are more pervasive than those of rural populations engaged in agriculture and fishing, sometimes in idyllic, usually coastal, environments. Indeed agriculture and fishing are the main sources of income and livelihood in the region, although the total value of mineral exports is greater than that from agriculture. Ironically, however, the value of food imports is generally greater than the revenue from all agricultural exports, and in some countries agricultural productivity per capita has fallen since the 1970s.

The causes of the lack of agricultural development include low world prices, infertile soils, poor management and even adverse weather. Land tenure systems sometimes constrain the productive use of land whilst the desire for a regular wage income ('fast money') and the perceived low status of agriculture discourage labour commitment. Agriculture is becoming less labour intensive, displacing people to find other employment. Such trends, stimulated and exacerbated by migration from rural areas, in turn dampen enthusiasm for agriculture, despite the growth of urban markets. In Palau and the Marshall Islands, agriculture has largely disappeared from all but the most remote places. There, and also in Tuvalu and Kiribati, there is virtually no local marketing of food. Urbanisation has contributed to the decline of agriculture, rather than its revitalisation.

In the Marshall Islands, for example, some agricultural systems have almost disintegrated in the face of imported foods, and despite subsidies for copra production – the sole viable economic resource – the returns are poor:

> The people of Majkin [Namu atoll] face a bleak future. The Marshall Islands government wants to encourage people to stay on their home islands, and thus reduce the urban pressure. But the resources on an outer island such as Namu cannot support any further population increase. The limit has been reached between supply and demand for food. Urgent attention is needed by the central government to replace coconut trees, increase the

> number of breadfruit trees, and to introduce an acceptable means of restricting population size. Otherwise outer island living will no longer be an idyllic option.
>
> (Pollock 1996: 10)

As expectations have increased, the focus of economic and social life has shifted from the rural peripheries to the urban centres. None the less, at least in what may be the exceptional case of Tamana atoll in southern Kiribati, the agricultural economy has grown rather than stagnated, since:

> People recognize that life in the urban centre has its detractions and that rural dwellers have somewhat more control over their lifestyle and living standards; they have the opportunity of being more self-reliant . . . In the face of uncertainty and limited resources, this is the best hope for a sustainable future.
>
> (Lawrence 1992: 55)

Prospects are rather better in Polynesia and Melanesia. As the experience of Tamana suggests, the attractions of rural life are in some part forced by the problems of urban life, rather than the merits of rural economies and societies. Agricultural life has an uncertain future.

Fisheries have enormous potential throughout this vast region of small land areas and large EEZs, yet the present potential of the seas is mainly limited to the exploitation of its extensive fisheries resources by distant-water fishing nations with limited benefits to the island states. Fisheries lease fees are an extremely important source of national income in several states, especially the Marshall Islands and FSM. Most processing is done outside the island states hence the fishing industry has usually contributed little to employment, especially in urban areas. Forestry is only important in PNG, the Solomon Islands and, to a much lesser extent, Fiji. Both in PNG and the Solomon Islands, the industry grew rapidly at the start of the 1990s, with a massive increase in investment and exploitation by Asian companies denied access to increasingly protected Asian forests. In both states the focus has been on log exports, rather than processed timber, and the island states – and particularly the landowners – have gained relatively little of the value of the timber, despite increased warnings from the international community that improved regulation was necessary. In neither country is the industry sustainable, little employment has been created, environmental degradation has ensued and the most accessible forests have been worked out.

Mining

Mining is the region's most important export sector, but its impact is very uneven. Almost all contemporary mineral exports come from PNG, though Fiji produces a modest amount of gold (and may develop the Namosi copper deposits); the Solomon Islands established its first modern mine on Gold Ridge in 1997, though this closed during the violence in 2000. Vanuatu also offers mining potential. Since the mid-1990s there has been a massive increase in interest in the extent of mineralisation around the 'Ring of Fire', running through the Asia-Pacific rim, with an increase in prospecting and the development of new recovery techniques enabling the exploitation of resources of low-grade ore in marginal locations. In the inter-war years some of the few towns in the Pacific – such as Vatukoula in Fiji, and Wau and Bulolo in PNG – were a direct result of mining activities.

In PNG mining had already passed agriculture as the major export before independence in 1975, after the Panguna (Bougainville) gold and copper mine had gone into production. In the 1980s there was some expansion and diversification with the opening of the Ok Tedi mine, an extraordinary gold rush at Mt Kare in one of the most remote parts of the highlands (see Figure 4.1) and the start of Misima gold mine. This brought economic growth and at least one new urban centre – Tabubil – into some of the least developed parts of the country. The Panguna mine closed in 1989, half-way through its working life, following clashes between local landowners and security forces, initially concerned over land alienation and environmental degradation, which eventually resulted in civil war and created a national economic crisis. However, the Porgera gold mine opened in the highlands in 1990, natural gas and oil were discovered in the Southern Highlands and resulted in the nation becoming an oil exporter in 1992, whilst construction of the Lihir gold mine began in 1996. The Lihir mine, which went into production in 1997, will be the largest mine in the southern hemisphere. Mining has utterly transformed the structure of the national economy, and in many respects distorted it, created a series of new towns (at Arawa and Panguna in Bougainville and Tabubil in Western Province) and reshaped the pattern of regional development (Connell 1997a). Nowhere else has economic development been so influenced by mining. Offshore minerals offer some prospects for future development in the eastern Pacific but their exploitation is currently uneconomic, and all countries, including PNG, are importers of energy at considerable expense.

Though mining has restructured the PNG economy and brought towns to remote areas, for an economic activity of such significance its impact on urbanisation has been surprisingly limited. In a few places mining towns did transform the landscape. Arawa, the prototype mining town, had grown from nothing to around 17,000 people at the time the mine closed; it was little different from similar towns elsewhere in the world: a somewhat sterile collection of houses of various kinds, with numbered sections rather than named and distinctive suburbs, and an extremely heterogeneous population. Despite steady growth in size and a greater diversity of functions Arawa never quite lost its aura of being a company town that was benefiting a distant rather than a local population. Other smaller and more recent mining towns are even more obviously company towns. Tabubil (Western Province) is similar but in a very different mountainous environment, grown beyond its 'artificial' start, with

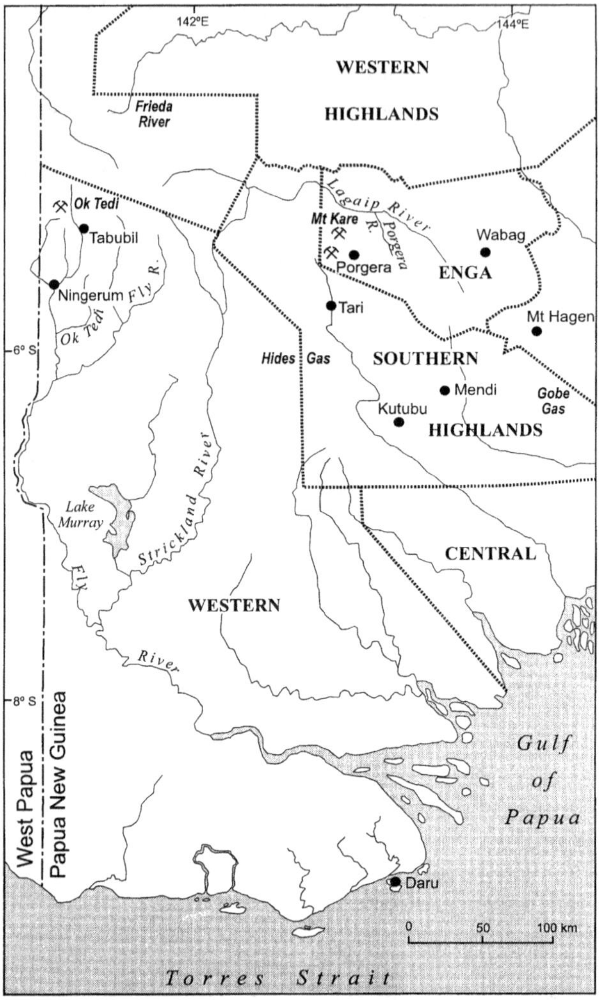

Figure 4.1 Mining towns in Highland Papua New Guinea

substantial migration from Oksapmin and elsewhere to the fringes of the town, in stark contrast to the contemporary formal layout. None the less,

> the cultural climate of Tabubil is decidedly colonial; the social hierarchy of the company and town is racialized, ethnically stratified, male dominated, and the superiority of expatriates is broadcast in crude and subtle ways. Though most workers in town are Papua New Guineans, the upper ranks are predominantly white, and nationals are regarded by the Europeans as a foreign race, alternately 'diverse', quaint and menacing.
>
> (Polier 1996: 3)

Porgera has become the last of the new mining towns. Much of the previous township was destroyed in 1991 by landslides, rendering it uninhabitable. New town construction followed. Spontaneous migration was even more visible there; in the five years before mining began there was a 50 per cent increase in the local population, mainly of Enga and Tari migrants, to the extent that landowners eventually sought company help in removing outsiders (Banks 1993: 323) and some squatters were forcibly removed in the development of a 'cordon sanitaire' around the mine site. Mining towns have only tenuous connections to the local region.

Urbanisation has often occurred at a distance from the mine sites. Because of the high cost of constructing facilities at Tabubil, itself 20 km from Ok Tedi, some mine support functions were located in Kiunga and Port Moresby, which also meant that some private urban costs were transferred to the public budget, and there was a speculative boom in house prices there (Pintz 1984: 161). Kiunga, too, received a boost from the growth of Ok Tedi, despite being 100 km from the mine. Away from the mining towns themselves, the most rapid mine-related growth has been at Mount Hagen, which has benefited from mining activities at Porgera, Mt Kare and Kutubu. Lae, at the end of the Highlands Highway, has also been boosted by the import and export demands of the mining boom, but Port Moresby, the base of many head offices, supply companies, ancillary services and transport ventures, has probably been the single greatest beneficiary of the urban impact of mining growth. However, the establishment of 'commuter mining' in PNG, through 'fly in–fly out' (FIFO) schemes based on Cairns in Queensland, reduced the local impact of mining. Since all the mining companies are transnational corporations, numerous specialised high-level operations attached to mining are undertaken overseas, and FIFO schemes expand this offshore activity. For mining companies commuter mining is more economic since it reduces the initial costs of town and infrastructure construction, some of which will be used for less than a decade, and substitutes recurrent costs. Both at Misima and Porgera there have been local demands for a mining town, rather than FIFO, because of the contribution to local economic growth. At Porgera landowners demanded that if the FIFO scheme continued, they should be compensated for the losses to the Porgera community (Connell 1997a: 151). Mining has boosted economic growth in northern Australia, at the expense of PNG, as there is less impact on the construction and service sector, with expatriate expenditure being largely transferred offshore. Through 'commuter mining' the mining sector is more obviously an enclave industry, without local linkages, especially near the smaller mines.

Manufacturing industry

Virtually throughout Oceania, industrial development has failed to go beyond the initial import-substitution phase, mainly associated with the production of such manufactured foods as bread and beer. Industrialisation is constrained by limited resources, the small and fragmented domestic market, inaccessibility and the impossibility of economies of scale; at least as important are high wages, especially in comparison with nearby Asia. The export of manufactured products remains a minor element in the South Pacific, and is unlikely to increase as free trade steadily replaces protection.

Much the most important exporter of manufactured goods, mainly textiles, has been Fiji, where the manufacturing sector grew rapidly in the post-coup years, following a series of new national incentives to international investors. The economic malaise that followed the 1987 coups led to radical restructuring of the economy, with the manufacturing sector seen as the panacea for economic ills. The garment industry gained prominence, resulting in the employment of many women, often in exploitative circumstances, with low wages and long hours, and without security or unionisation (Slatter 1987; Lateef 1990; Emberson-Bain 1994a). In the 1990s the period of post-coup growth ended and by the mid-1990s the manufacturing sector in Fiji had stagnated. The garment industry remained relatively strong, employing half of all workers in the industrial sector (about 18,000 people) and with exports valued at F$185 million (US$120 million). Nevertheless, industrial employment increased by no more than about 1 per cent per annum and most jobs were unskilled. Employers attributed this to slow economic growth, low demand for their products, the poor quality of the labour force (particularly low productivity, high absenteeism and the lack of skilled workers), insufficient finance, ill-considered government policies and restricted access to raw materials (UNDP 1997: 18). The 2000 coup destabilised and diluted the already stagnant industrial sector. Fiji, the most successful industrial economy in the Pacific, has found it difficult to sustain manufacturing growth.

The largest country in the region, PNG, has been even more unsuccessful in establishing industrialisation. Very high wages, lack of development capital and skilled personnel, a small, fragmented and low-income domestic market, and high insurance and utility costs have all discouraged industrial development. The industrial sector (excluding mining) contributes about 13 per cent of GDP and only 8 per cent of formal employment. Industrialisation in PNG remains focused on import substitution: around breweries and meat and fish canneries, none of which is economic, though it has contributed to employment in Madang and Lae (Connell 1997a: 203–4). High wage rates, almost twice those of the industrialising states of Asia, whilst discouraging growth, have also resulted in over 90 per cent of the industrial workforce being men, a very different situation from Asia and everywhere else in the Pacific region. Wages are high because of the cost of extended family obligations, the security risks and costs of living in urban areas and the small size of the skilled national workforce. Land tenure and land administration procedures have made it difficult and expensive to acquire land for business ventures. Land constraints have also contributed to high accommodation costs. An industrial estate policy was introduced in 1988 to minimise land issues, with the first industrial estate at Lae. However, industrial development failed to follow, despite clothing manufacturers being given special import concessions in Australia. By the early

1990s there was just one mass production garment factory – its lonely presence symbolic of the failure of industrialisation – and the estate barely grew subsequently (Chen and Yala 2000).

In the Solomon Islands and Vanuatu industrial development has been even less evident than in PNG, despite the emergence of a fish processing industry in the former and some import-substitution activities. In both countries ambivalence towards foreign investment, bureaucratic requirements that are 'excessively cumbersome and almost designed to ward off potential investors' (Asian Development Bank 1997: 86) and political instability have discouraged industrialisation. Industrial development has been more successful in the smaller Polynesian states of Tonga and Samoa. In Tonga manufacturing grew after 1980 following the establishment of a Small Industries Centre in Nuku'alofa. A number of foreign-owned enterprises began operation through the Pacific Islands Industrial Development Scheme (PIIDS) and other incentives. The principal industrial development concerned foods and was primarily orientated towards import substitution. New industries were involved in the export of leather goods, knitwear, quilts and small amounts of industrial machinery, so that the Small Industries Centre had grown to employ more than 240 people by 1992. The steady success of this centre resulted in the opening of a second centre at Neiafu in Vava'u in 1990. However, despite strong export growth in the 1980s, which led to increased employment and a 10 per cent contribution to GDP by the end of the decade, as elsewhere in the Pacific, the industrial sector was faltering by the 1990s (Connell and Lea 1995: 42). By 1993 manufactured exports were at their lowest levels since the mid-1980s. Factories laid off workers and experienced temporary closures, creating a noticeable growth of urban unemployment in the capital, and by 2000 there was little left of the Small Industries Centre, other than a brewery. Until quite recently manufacturing in Samoa was essentially small-scale and orientated to import substitution but, in the 1980s, the Vailima brewery became the country's biggest industry (and accounted for 12 per cent of the value of exports in the early 1990s). Cigarettes, manufactured in Apia, accounted for 7 per cent of exports. The Japanese Yazaki company in suburban Vaitele was a major new development in 1991 to manufacture car seat belts and electrical wiring systems, in a plant that employs 1,500 people, the single biggest industry in the Pacific. However, it has experienced problems, with local complaints over low wages (the principal national benefit), rapid labour turnover and high rates of absenteeism, hence its long-term future is uncertain. No other comparable industries have located in Polynesia.

Constraints on the establishment of manufacturing industry in the central Pacific are considerable and include the negative characteristics that discourage most forms of modern sector development there and elsewhere in the region: difficulties with land tenure, few appropriate skills, poor infrastructure, few domestic resources, a lack of credit, and a regulatory environment that discourages private investment. In Kiribati, the Marshall Islands and Palau there is no manufacturing beyond some trivial food processing and coconut oil and soap production. The Marshall Islands established a brewery in Majuro in 1996 and a garment factory similar to that operating in Yap with Chinese assistance. In the FSM manufacturing was similarly low key and exports minimal until the establishment in 1988 of the foreign-owned garment factory in Yap; the factory imports raw materials and exports its produce to the USA, taking advantage of duty-free status. It formerly employed 300 Sri Lankan women but the workforce subsequently became almost wholly Chinese on exceptionally low wages. In Kiribati manufacturing employs some 600, accounts for about 2 per cent of GDP and mainly involves boat building and handicraft production. Experience of manufacturing in Micronesia suggests that without unusual developments of the kind described for Yap, the prospects for expansion are slight.

Despite often strenuous efforts no countries have achieved great success with industrialisation and even early successes have not been sustained over time. The most successful manufacturing industry in the region, the garments industry of Fiji which employs around 9,000 people, has stagnated (and was in decline in 2000 as Australian subsidies ended, Fiji garment manufacturers faced severe competition from Asia, trade preferences disappeared and political instability intensified). It is symptomatic of the problems of achieving industrialisation that the most important manufactured goods exported from some states, and especially the smallest, are postage stamps.

Financial services

Recently there has been some expansion into financial services, most apparent in the emergence of a tax haven and finance centre in Vanuatu. Vanuatu's financial services industry began in 1980, expanded substantially in the 1980s but grew only slowly since then because of corruption issues. By 1989 over 1,600 companies were registered there, 380 people were employed by the finance centre and it has annually contributed about 10 per cent of Vanuatu's GDP (Connell and Pritchard 1990), a situation more or less maintained subsequently (Rawlings 1999b). Although the number of finance companies had increased to over 2,400 by 1995, employment had failed to increase and the contribution to the GDP was no greater (Asian Development Bank 1997: 63–4). In the 1980s other island states – Tonga, Nauru and the Cook Islands – also established tax havens, none successfully. Both Vanuatu and the Marshall Islands have flag of convenience fleets, the latter being ninth largest in the world, which have generated some income but little employment. Few states will succeed in the world of finance but the change is indicative of both the growing diversity of economic development in the region and the transition from rural primary production to the urban service sector.

The external context

Foreign financial assistance is crucial to economic life for virtually all South Pacific countries. Except for Fiji and to a lesser extent PNG, all countries have external assistance levels which account for significant proportions of both their GDP and their total governmental expenditure. Fiji's figures are less than 5 per cent of GDP and 10 per cent of government expenditure. By contrast, Tuvalu's level of aid as a percentage of both indicators exceeded 100 per cent before the 1988 decision to establish a special trust fund. Per capita aid flows are exceptionally high by global standards, as the island states have benefited from the general 'small-country bias' that marks the aid-giving policies of most donor nations. This bias, resulting from their economic, administrative and locational advantages, was earlier reinforced by political and strategic considerations, though the declining strategic significance of Oceania has produced aid fatigue and a reduction in aid flows from several donor nations, particularly the United Kingdom and the United States. Overseas aid to Pacific states declined by 2 per cent per year in real terms over the 1983–93 period, and even more rapidly in subsequent years. Dependence on high levels of aid is increasingly precarious, especially for the Compact states of Micronesia, where aid levels are exceptionally high but projected to fall sharply. Moreover, dependence on aid has made it relatively easy for donors and international agencies such as the ADB and the IMF to push agendas of deregulation, privatisation and structural adjustment.

A high proportion of aid goes into infrastructure provision, and this 'technological bias' has been criticised because of its urban focus, limited spread effects and high maintenance costs (Connell 1986: 72) alongside its discouragement of the private sector by encouraging state bureaucratic growth (Schoeffel 1996: 135–6). If aid has contributed little to economic development in Pacific states it has produced welfare benefits, especially in the smaller states. Because of the urban focus of aid delivery and expenditure, which has emphasised the bureaucratic component of MIRAB economies and the expansion of infrastructure, aid has influenced the extent and form of urbanisation. Island states are likely to remain aid-dependent for the foreseeable future, though access to aid may further decline.

The Pacific's balance of trade runs strongly against the region. Though the better resource-endowed states of Melanesia enjoy a more favourable trading relationship with their partners, only PNG, Fiji and the Solomon Islands have had trade surpluses since the 1980s, primarily because of the expansion of minerals and timber exports. (For the Solomon Islands this is unsustainable.) Minerals are the region's main exports, representing about half the value of all exports and more than 70 per cent of the value of PNG's exports, which since 1991 have included oil and should soon include natural gas. Agriculture provides the second major category of exports and, unlike minerals, virtually all countries in the region are represented. Even here there are significant variations. PNG's substantial coffee and cocoa industries alone contribute about 15 per cent of the region's total domestic exports. Fiji's annual export of sugar returned about 7 per cent of the total. In Fiji itself sugar exports represent more than half the value of all exports. The structure of imports remains the converse of that of exports. Machinery and transport equipment are the major imports, partly as a result of the growth of the mining industry. The weaker, resource-poor countries devote a much greater proportion of their import expenditure to foods than do the larger countries. Import growth is a major economic feature of island economies, emphasising worsening trade balances. The legacy of the region's colonial experience remains evident in trade patterns. The metropolitan states which dominated the region from the Second World War, especially Australia and New Zealand, remain significant economic influences, despite newer Asian links. Regional integration of trade is restricted because of the minimal complementarity of production and poor transport connections.

Tourism

The idyllic image of the South Seas has made the region a tourist mecca in the public mind. But, as with so many other sectors, the economic benefits of tourism are limited to few countries and, overall, the tourism industry is less important than either the tuna industry or mining. None of the independent states has achieved the success of Guam or Saipan (Northern Marianas), which have captured a large proportion of Asian visitors, and for many years there was reluctance to embrace tourism because of its perceived negative cultural impact. Fiji has been much the most successful. Although the 1987 coups led to hotel occupancy levels reaching only 10 per cent of their capacity, by 1990 tourist numbers had returned to pre-coup levels, and they subsequently increased further, until the 2000 coup again led to a collapse. Tourism is still the major source of economic growth in Fiji. In 1999, it had 410,000 visitors, many more than the 55,000 who visited Vanuatu, the second most important tourist destination in the region. Political disturbances, cyclones, airline policies and changes in hotel ownership have all discouraged steady growth. The Solomon Islands had intermittent success, experiencing land tenure difficulties with some hotel sites, and the industry collapsed in 2000. Nevertheless all these smaller Melanesian states have been relatively successful compared with PNG, where high costs and insecurity have resulted in the stagnation and decline of the industry, despite numerous attempts at promotion.

Palau has a growing tourist industry, which is the most important source of national income (after aid) with most tourists being from Asia. Samoa and Tonga also have small tourist industries; though they contribute relatively little to the economy and stimulate only a small proportion of employment, there are considerable prospects for growth in comparison with other sectors of the economy. Throughout the Pacific tourist arrivals increased by 40 per cent in the first half of the 1990s, but this rate of growth was then slowing. Ecotourism has not yet emerged, and most tourist facilities are on the fringes of urban centres, except in Fiji. Isolation, poor communications, little promotion, instability and intervening opportunities have limited the development potential of an industry in which Oceania might otherwise have a comparative advantage.

Development

According to United Nations classifications, four of the region's economies – Kiribati, Tuvalu, Samoa and Vanuatu – are least-developed countries. The other Pacific island states are considered developing nations, yet all had human development indices that were worsening in the 1990s (UNDP 1999: 15). The larger Melanesian states have greater potential for economic growth but social development is weak. Rates of economic growth that are everywhere rarely above those of population growth have restricted the capacity of governments to finance service provision. However, social indicators for even the least-developed countries compare favourably with those in most other regions of the world. Life expectancies are lowest in the Melanesian states but, except in PNG, are still increasing slowly. Despite most countries, outside Melanesia, having adult literacy rates of over 90 per cent (Table 4.3), the education system demonstrates widespread inefficiencies and has largely failed to produce a skilled labour force. Low skill levels are endemic across the Pacific, in terms of modern economic activity, and constrain economic growth and improved human development (UNCHS and UNDP 1994: 22). The motivation of governments to achieve real change is weak. For both education and health, government expenditure is declining in real terms in most states.

With the exception of PNG, the Solomon Islands and Vanuatu, which are regarded by the UNDP as countries of 'low human development', development in the Pacific compares reasonably well with that in other developing countries, though this may be more difficult to maintain in the future. Cultural diversity, geographical fragmentation, rapid population growth and a very dispersed population (in a region of islands and challenging topography) and the late development of modern infrastructure have emphasised the problems of achieving development in the Melanesian states. In PNG, access to doctors, health posts and other rural services has worsened since independence, as services are increasingly concentrated in urban areas (Connell 1997b) and a similar trend is becoming apparent elsewhere.

Development plans in most countries initially stressed the preservation of traditional cultures and the necessity to achieve a greater degree of self-reliance, at least in terms of balancing incomes and expenditure. By the 1990s most countries had hesitantly adopted economic management policies which endorsed deregulation and a reduction in government intervention and many governments had instituted economic reform programmes that related to these aims. Restructuring was the externally imposed response to what the World Bank had referred to as the 'Pacific paradox': favourable natural and human resource endowments (though, at best, these were scarcely ubiquitous), high aid levels and generally sound economic management, but no evidence of growth in per capita incomes (World Bank 1993c; cf. Knapman and Saldanha 1999). However, in most countries macro- and micro-economic reforms are incomplete and the processes of restructuring and adjustment have a long way to go; there has been enormous reluctance to restructure and deregulate, notably in PNG,

Table 4.3 Human development indicators in the Pacific

Country	*Human Development Index (HDI)*[a]	*Literacy rate*			*Mean years of schooling 1991*	*Gross school enrolment*		*Access to safe water 1991 (%)*
		Adult (15+) (%)	*Male (%)*	*Female (%)*		*Males*	*Females*	
FSM	0.57	83	85	81	7.6	71	71	30
Fiji	0.67	93	95	91	6.8	80	83	92
Kiribati	0.52	93	94	92	6.1	67	69	65
Marshall Islands	0.56	92	93	92	8.5	72	72	50
Palau	0.86	98	98	97	9.6	83	84	—
PNG	0.31	63	71	56	2.1	30	27	23
Samoa	0.59	82	83	82	9.1	84	88	70
Solomon Islands	0.37	70	65	55	2.8	38	31	61
Tonga	0.65	98	98	98	7.1	82	84	100
Tuvalu	0.58	98	98	98	6.8	74	75	100
Vanuatu	0.43	76	78	74	4.0	60	55	87

Sources: Forum Secretariat (1996: 70); Schoeffel (1996); Secretariat of the Pacific Community (2001)

Note

a HDI is a composite index which takes account of life expectancy, educational attainment and income. The Pacific HDI is not directly comparable to the global HDI figures published on a regular basis by UNDP.

in the face of Asian Development Bank, World Bank and donor country pressure (Connell 1997a; Sutherland 2000). Despite some deregulation many states remain characterised by the public sector, in terms of both employment and its involvement in the kinds of public utilities that are privatised elsewhere. Few of the region's states have the political strength to ensure that an adequate package of economic measures is implemented over the longer term, hence public sector reform is difficult. Moreover, social obligations and traditional cultural practices, of great importance to island peoples and governments, can constrain modern economic development, especially through the relationship between land tenure and rural development.

Employment, unemployment and poverty

In much of the region, agricultural activities dominate employment, usually combining subsistence, cash cropping and fisheries. Outside agriculture, most employment in the region is in the service sector and especially in government. The public sector grew rapidly around independence in most states, but growth has either ended or has actually declined, as in FSM and the Marshall Islands, as aid too declined. Economic recession at the start of the 1980s severely restricted the growth of new wage employment opportunities and in many countries there has been a growing problem of urban and especially youth unemployment.

Female participation in the formal workforce has increased; however, especially in Melanesia, this remains slight. Urban occupations where women have traditionally been employed – the retail sector, manufacturing and the public sector (including education and health) – have been relatively stable, and women have become more likely to gain an education. Even so, women experience discrimination, 'glass ceilings' prevent promotion and many are prevented by traditional values, including high fertility, from more active participation in wage employment, with greater time spent on child-rearing and domestic work (see, for example, M. Ward 1995). In Fiji, where women made up 33 per cent of the paid workforce in 1997, many occupations are effectively reserved for men, and employment opportunities for women are segregated in a small corner of the labour market (Bryant 1993: 46–8; UNDP 1997: 23). In Levuka, Fiji, women are predominantly employed in the large tuna cannery because they are said to be 'more gentle with the fish, and less likely to be bored' (personal communication 2000; cf. Emberson-Bain 1994a). At the Yazaki automotive parts plant in Apia women are absolutely dominant on the shop-floor, earning the minimum wage for laborious shift work in component assembly; here too they are said to be 'better than men in this kind of work' (personal communication 2000). Gender stereotypes abound. Whilst traditional gender roles restrict women's participation in the formal sector, such traditions often change when new income-earning opportunities exist.

Because of the agricultural focus and the mainly rural population distribution, formal unemployment levels are low but growing. Increased levels of unemployment accompany urbanisation, with a growing strain on traditional social support networks. In the urban formal employment sector wage and salary levels and employment creation are only very partially determined by the workings of a fully competitive labour market. Formal sector employment is concentrated in urban areas and a significant proportion of that employment – almost all in the case of Tuvalu – is located in the public sector (Table 4.2). In most towns the employment structure is heavily dominated by privileged tertiary sector activities where incomes are not related to productivity but to institutional factors. While average formal sector income levels are somewhat higher than that at which full employment could result, migrants continue to be attracted to towns as long as they perceive some chance of finding a job in the formal sector. Education, even in rural areas, prepares students for employment outside agriculture and leads them to aspire towards urban white-collar employment. Taking the risk of a prolonged period of unemployment before finding a formal sector job remains rational.

The employment structure of the towns and cities is tripartite: an administrative and service sector (which is characteristic of all towns, was the principal reason for the establishment of many and is often almost the only real contemporary economic function of the smaller towns); a small manufacturing sector; and the 'informal' sector. The early growth of most Pacific towns was linked to the emergence of administration, and administrative employment has gained high status throughout the region. More importantly, public sector wages are often substantially above average wage levels, especially in the former TTPI states; this has tended to set a benchmark for those in the private sector and has discouraged its development.

Unemployment is usually considered to be worse in the informal settlements, especially those of recent origin, yet what Jackson (1976d: 49) recorded of PNG towns a quarter of a century ago remains true: 'settlements are frequently the homes of people less transitory, no more unemployed and just as urbanized as other sectors of the population'. In several PNG settlements in the 1970s, not only were a majority of adults employed, but public servants, entrepreneurs and even MPs lived there (Goddard 2001: 10). In Blacksands settlement, Port Vila, most people were employed in the city, including senior government positions, alongside work in the private sector or as a pastor. The service sector dominated employment through drivers, domestic workers, construction workers, gardeners or shop assistants; increasingly women took on 'non-traditional' roles, such as working in stores, and became the main income earner in some households. Employment was linked to skills (and ages) rather than kinship ties. The broad structure of employment in the settlement was similar to that in Vila as a whole (Mecartney 2001).

The transition from colonial towns to Pacific cities has accompanied the diversification of the urban economy, despite the lack of an 'informal sector', even in the large cities of PNG, where the combination of migration, growing urban permanence, few new urban employment opportunities and the lack of industrialisation might have been expected to have resulted in the emergence

Plate 4.1 Bottle collecting, Boroko, Port Moresby, Papua New Guinea. © John Connell

of such a sector (where wages and conditions of work are unregulated). However, there has long been an 'almost total absence of handicraft workers and the small-scale traders, shopkeepers and service-repair establishments that provide such a large volume of employment, especially urban employment, in many other countries' (World Bank 1978: 55). The most visible elements of the informal sector are crime (and its converse, security provision), betel nut selling and bottle and can collecting (all particularly evident in Port Moresby), alongside house-building, food-selling, transport, newspaper selling, window-cleaning, prostitution and gardening. The informal sector overall provides incomes and employment for a very large number of people, usually at wages and incomes below those in the formal sector.

The relative absence of informal sector activities in the Pacific is partly due to restrictive legislation. There is a 'common and pervasive bias against small-scale industrial ventures that are carried out within the purview of the so-called informal sector' (Fairbairn 1992: 24). Whilst, in most countries, a range of informal economic activities exist that are not subject to the rules and regulations which govern businesses in the formal sector, in PNG they receive no official recognition or promotional assistance, are over-regulated or simply disrupted. This is especially true of betel nut selling and food selling, where

Plate 4.2 Attitudes to the informal sector, Papua New Guinea. This cartoon contrasts the regular harassment of food and second-hand clothes sellers in several Port Moresby markets with the relative freedom given to prostitutes and beggars.

Source: *The Independent* (Port Moresby), 7 September 2000; reprinted with permission of *The Independent*, Papua New Guinea

tough colonial legislation remains in place. At the start of the 1990s there were considerable efforts to close down and remove illegal 'tuckershops' in Port Moresby, which had no licences and were operating on state land. From time to time there have been health bans on the selling of cooked food in streets and markets, and betel nut vendors are routinely harassed, have their goods stolen and confiscated or are banned entirely. In 1993 the National Capital Development Commission (NCDC) brought in a law to license street bottle collectors and restrict their trading to 8 a.m. to 5 p.m. on weekdays and 9 a.m. to 5 p.m. on weekends, a law virtually impossible to administer effectively with or without harassment. Other reasons for the limited significance of the informal sector throughout the Pacific are low consumer demand for some services, the relative affluence of the rural sector (which takes the burden off some of the urban unemployed) and, perhaps, an absence of skills in some areas (such as shoe repair or sandal manufacture) where demand is only emerging.

Where there is no restrictive legislation informal sector activities have grown rapidly. This is particularly true of vehicle transport, since there is no state or municipal public transport sector in any town in the South Pacific, apart from Suva. In the larger towns, notably Port Moresby, there are hundreds of mini-buses, many owned by wealthy individuals and business groups from the highlands. Driving is regarded as a highly prestigious job and wages may be above those in skilled formal sector activities.

The informal sector is most evident in the extent and significance of produce markets, long a feature of Pacific towns (Brookfield and Brown Glick 1969), which have become larger, more numerous and more specialised in post-independence years and contribute to significant rural income generation. They have also spawned the emergence of market entrepreneurs and 'middle women', especially in the larger cities of PNG (Connell 1997a). Vendors in urban markets are part of the urban informal sector since, other than in their payments for renting market stalls, their presence is unregulated and sometimes intermittent. Relative to other parts of the world:

> the paradox of the informal sector in Papua New Guinea's towns is that so much of it is based on an abundance of natural resources, such as unused peri-urban land and fisheries, whereas in other countries the scarcity of such resources forces many urban dwellers into service and trading activities more typically associated with the informal sector.
>
> (Conroy 1982: 16)

In the early 1980s there were around 1,000 market vendors in Port Moresby on peak Saturdays (Walsh 1982b) and other towns, such as Suva, Honiara and Port Vila (Bathgate 1978; McGee *et al.* 1980), also had large numbers. A decade later the number of formal markets, and informal market places (often on the edge of PNG towns to avoid regulations about betel-juice spitting), had increased, the number of vendors become much greater and the volume and range of goods considerable. Gambling in market places has become a common 'service' sector activity, though manufactured products are still rare. Markets play a very much greater role in every urban economy than they did at the time of independence.

Only the smaller towns, such as Majuro, Tarawa, Funafuti and Koror, where there is little vegetable production, are effectively without markets. By contrast, Apia and Nuku'alofa have large, diversified and distinctly different markets, a function of international migration and the flow of goods as remittances. By the late 1980s the central Nuku'alofa fruit and vegetable market at Talamahu was surrounded by an untidy collection of individual vendors (and some stalls) almost entirely selling second-hand clothes: the first sign of a 'flea market' later selling 'a wide variety of second-hand goods including clothing, furniture, books, cassettes, machinery and even the kitchen sink. Many of the items sold enter Tonga without paying duty, and therefore have an edge in competition with local traders' (James 1991: 7; see also Brown and Connell 1993). In 1994 the flea market had been decentralised to a site on the main road out of

Nuku'alofa, in theory to make way for a replanting project. In practice, the formal sector, through the Chamber of Commerce, had raised objections to the strength of competition. Outside the flea market there were also individual food vendors along the main Taufa'ahau Road, especially in the evenings; some were recent migrants, either jobless or able to earn more from this sector, mainly from Ha'apai. In Samoa too, in the 1990s, there were similar and growing demands from business to decentralise competing retail activities and bring them under a tighter and more restrictive fiscal regime. It was recorded in the press that 'any clean-up effort of Apia cannot succeed if the vendors doing business along the streets are not moved to a location. This is where most of the filth in the town emanates from' (*Samoa Observer*, 24 January 1992), a conclusion that was without much evidence. In the market sector of the economy at least, the informal sector has become of considerable local importance, although, in Tonga, success brought conflict with the formal sector. Throughout the region there has been little attempt to provide protection and space for the informal sector despite its positive impact on both incomes and employment.

Urban informal sector activities are usually much less visible than markets and transport. Small-scale manufacturing activities include cooking, tailoring and handicraft production, and many tiny businesses operate from homes – producing and selling individual cigarettes, operating *kava*, *sakau* and 'grog' shops and bars or, beyond the law, engaging in prostitution or marijuana sales (Bryant 1993: 78; UNDP 1997: 28) – demonstrating that the principles of entrepreneurialism and capitalism are not confined to the most formal urban echelons. Many such businesses may be small and short lived, but they generate considerable employment, mainly of youths and often of women. A few have grown, partly by engaging in black market operations, by setting up pool (billiards) tables or through particular entrepreneurial skills in good locations. Despite a high level of unemployment less than 10 per cent of income earners in Blacksands (Vila) were engaged in informal sector activities, mainly because the market was small and competition was considerable, and lack of land, skills or licences prevented success. Income-earning activities included handicraft production, small-scale trading (of fish and garden produce), small-scale production (mostly cooking, tailoring and carpentry) and domestic work; some activities, such as roasting and selling peanuts, were linked back to home island suppliers. Work was invariably intermittent, seasonal and poorly rewarded (Mecartney 2001). As in the market sector, women predominate for several reasons: gender discrimination in the formal sector; limited education and skills; and the ability to combine informal income-generating with a domestic role.

If markets and small businesses are the positive manifestation of the informal sector through their contribution to incomes, employment and nutrition, the converse is crime and prostitution, increasing components of urban employment and life styles. The oldest trade is now 'booming in many Pacific island cities and towns. It is a flourishing industry which evades income tax, brings in foreign exchange, but may prove to be socially and economically costly' (Tara 1996: 22), especially in the capital cities of Melanesia. In Honiara the number of *dugongs* (prostitutes) increased rapidly in the 1990s, taking in school dropouts, unemployed students and some low-income employed, because of domestic problems or the financial burden of a rapidly increasing cost of living. (In the mid-1990s the inflation rate was over 10 per cent and only 500 jobs per year were being created for 10,000 school leavers.) In Fiji prostitutes are similarly motivated by economic pressures caused by unemployment, divorce, desertion, the failure of men to pay maintenance for their children and lack of assistance from their extended family (Naibavu and Schutz 1986; Government of Fiji and UNDP 1997: 66). Prostitution has grown dramatically in PNG, especially in Port Moresby, based around hotels and discos. Prostitutes are from settlements and urban villages, from 'broken homes' or very large families, single mothers or with unemployed husbands. Potential earnings were substantially above those in the lower income levels of the formal sector, with prices ranging from K20 to K200 at the end of the 1980s. At the start of the 1990s prostitution in Port Moresby was reported to involve over 200 women, including some Asian migrants, and has grown steadily since then, not only in the capital city but even in quite small urban centres such as Daru (Hammar 1992). In 1995 some 38 per cent of otherwise unemployed women were reported to be working as prostitutes – suggesting that there were well over a thousand in the capital – earning more than most employed women (Levantis 1997b). In Fiji the economic impact is also significant:

> prostitution is a fully localised industry which gives employment to unskilled female workers for most of whom no other jobs are available. It requires no investment of foreign capital, yet it brings in large amounts of foreign exchange with a minimum of leakage back overseas.
>
> (Naibavu and Schutz 1986: 98)

The almost universal rise of prostitution in the Pacific is symptomatic of the problems of limited economic growth, unemployment, the social costs of urbanisation, the decline of traditional social control mechanisms and, not least, potentially high earnings. It poses obvious health problems, including the rapid rise in HIV/AIDS cases in PNG.

Crime, and the resultant acquisition of goods and wealth, is also an informal sector activity, above all in PNG. In Port Moresby a 1995 survey revealed that a staggering 69 per cent of males, especially youths, who considered themselves to be unemployed were actually earning a living through crime (which translates to 19 per cent of the entire male urban workforce), and that the incomes obtained through this were above those in the formal sector. Hence few of the euphemistically titled *raskols* were actively seeking formal employment, for which they were largely without appropriate skills. Significantly, '*raskols* are doing what the government is unable to do and are redistributing income to the less fortunate who are unable to find formal employment' but at the cost of 'stifling any potential growth in output or employment', thus creating a 'spiral of disaster' (Levantis 1997b, 2000). The widespread significance of gangs points to their considerable organisation: 'Commercialism, *bisnis*, appears to be driving and transforming virtually all structures of Rascal organisation' (Nibbrig 1992: 119), with breaking and entering being central, sometimes to meet international orders. The escalation of crime in the 1990s in Port Moresby took it to a situation even worse than that of such notorious international cities as Rio de Janeiro, Kampala and Johannesburg, following a twenty-fold increase in crime from 1970–90 (Levantis 2000). In other large cities, notably Suva and Honiara (see, for example, Keesing 1994), crime is significant enough to have produced widespread concern. Crime has, inevitably, contributed to the rise of informal security services, with walls and chain fences, burglar alarms, large dogs and guards being new elements in the urban landscape. Begging has gradually become more significant, though less visible, importuning and organised than in many Asian cities. Everywhere the informal sector has grown, but most dramatically and painfully in Melanesia, becoming more visible, more significant for income generation, more organised and regulated, but usually poorly understood and rarely officially supported.

Declining job opportunities in urban areas have necessarily emphasised the difference between those with jobs and those without jobs. Urban unemployment varies significantly between different groups within towns, sometimes according to the time period of movement to the city. Thus there are differences between the relatively unemployed highlanders and employed coastals in Port Moresby (Levine and Levine 1979: 37–8), as a result of both discrimination and the superior education of many lowlanders. This kind of variation in urban unemployment emphasises existing inequalities in rural income, as migrants increasingly come from the more impoverished rural areas. However, the definition and measurement of unemployment is difficult (because of the significance of the informal sector and the continued importance of semi-subsistence activities) and few adequate studies exist. However, overt unemployment occurs in all urban areas and its extent is increasing. In some contexts this has long been apparent; in Vanuatu in 1969, 'all evidence suggests that there is presently a surplus of unskilled labour at least in the area around Port Vila' (Metcalf 1969: 14). Since then there has been much evidence of growing urban unemployment throughout the region. Though formal surveys are few, the visible evidence alone is striking.

Early studies of urban unemployment in PNG in 1973–4 revealed that those out of work tended to come mainly from a small number of relatively poor areas hence:

> Urban unemployment in Papua New Guinea was mainly a phenomenon of the relatively young, the less educated and the more recent arrivals in town. It seems to have been quantitatively most important amongst people from districts very close to the town, and from more distant districts that are poor in services and income-earning opportunities and yet linked into the country's main communications and transport networks.
>
> (Garnaut *et al*. 1977: 56)

Moreover migrants from such highland provinces as Simbu (Chimbu) were prepared to endure much lower wage levels (and employment rates) than migrants from other areas, an indication of the problems of rural development in Simbu. Relatively few migrants had then left very poor areas of PNG and, with the exception of some migrants from Gulf province, few migrant groups had produced many children in town. Thus immediately prior to independence evidence of unemployment had only just emerged. During the 1990s, and especially from 1995 to 2000, unemployment and permanent migration increased together. At the time of the 1990 census the overall urban unemployment rate in PNG was recorded at 29 per cent but, because of the significance of the informal sector, this was a substantial overestimate (Levantis 1997b: 79). A third of adults in the capital were officially searching for formal sector employment. A substantial number of urban households were without urban wage or salary workers: in Port Moresby and Madang, even in 1983, a quarter of all households in many settlements were without any employed member; many were dependent on gifts and transfers from others (McGavin 1991: 52–7) or from more intermittent earnings in the informal sector. Evidence of increased unemployment exists in the number of jobless youths in the cities, and in the prevalence on gates, especially of factories, of signs stating *Sori Nogat Wok* (Sorry, No Work), visible demonstration that demand for work exceeds supply but also that urban wages are not too low (ibid.: 156).

Throughout the region unemployment is particularly high amongst youths (Bryant 1993: 46). In Nuku'alofa the overall unemployment rate in 1986 was 11 per cent (compared with 8 per cent for Tonga as a whole) but for youths aged between 15 and 19 it was as high as 28 per cent and the female rate in this age group was more than 50 per cent, an indication of the growing presence of youth unemployment despite substantial emigration. Moreover, inadequate access to employment, land and credit have led to increased levels of unemployment in both the small Vava'u town of Neiafu and in Nuku'alofa (Gailey 1992) and this, in turn, has stimulated emigration. Much the same is true elsewhere, as in Samoa, where there is a growing number of unemployed secondary school graduates. There are some indications that a rising urban crime rate is one result of this and there are few programmes either to induce some of the unemployed to return to the rural areas or to develop appropriate employment programmes (Miles *et al.* 1992; UNDP 1993: 55). In terms of social and economic development in Samoa: 'Youth residing in urban Apia were singled out as having special needs. In Apia, social support and social control have loosened and productive employment opportunities are restricted by limited access to agricultural lands' (UNDP 1993: 59).

Youth unemployment is also high in Fiji, and again especially amongst those with poor educational qualifications in urban areas. It has contributed to growing urban crime problems and substance abuse:

> Teenage alcohol abuse is commonplace in Fiji and a contributing factor in many juvenile crimes. . . . Marijuana smoking among male youths is now

> common despite mandatory jail sentences for drug use. Some youths reportedly smoke marijuana all day, becoming unresponsive and inclined to unpredictable behaviour, violence and other anti-social behaviour.
>
> (Government of Fiji and UNDP 1997: 65)

Young men also make up the majority of the prison population. In Vanuatu as well, youth unemployment is increasing, with many 'killing time' in public areas; such youths have acquired the name SPR or *sperem publik rod* (literally 'roaming the streets'). In PNG they are *pasindia* (literally 'passengers') and in Honiara, the similar *liu*. In Blacksands (Vila) about 10 per cent of the population were unemployed, yet working lives were generally short, hedged in by insecurity, poor conditions and low wages (hence workers simply drifted out of work, often as a result of injury or illness, or left out of frustration), made more difficult by limited skills and little job creation. Employment aspirations were far from being satisfied (Mecartney 2001). Broadly similar conditions occur throughout the region, but are much less evident and undocumented in the smaller states.

Despite the significance of urban employment and unemployment, there are virtually no data on the particular significance of the urban economies and their relationship to national and regional development. The sectoral distribution of urban unemployment and, more significantly, the contribution of the urban economy to national development are virtually unknown. In Samoa the UNDP has recommended that development assistance be targeted towards economically disadvantaged or vulnerable households, some of whom were thought to be urban migrants, but wryly concluded 'there is no data base by which to identify disadvantaged or vulnerable households' (UNDP 1993: 10). The same could be said of most other countries.

There is little formal knowledge of the extent of urban poverty or welfare problems, but clear indications that the nutritional and health status is often worse in urban areas. In Samoa in 1979 the extent of malnutrition in Apia was estimated to be three times that of rural areas. In 1992 half of all malnutrition referrals came from Apia or suburban areas. Non-communicable diseases are also more significant in Apia where their extent is twice that in rural areas (ibid.: 45–6). This latter situation is also true of Nuku'alofa (Finau *et al.* 1987) but there is less evidence of malnutrition. In both these countries the extent of squatter settlements is very limited, whereas in Melanesia, where settlements are more significant, unemployment rates are higher, as are the visible and measured indicators of nutritional and health status.

Absolute poverty is not generally apparent in Oceania; however, some households are poor in the sense that they do not have enough food, clean water or access to an adequate house or basic education (Bryant 1993: 13) but who and where these households are is little known. Certainly poverty is more visible in urban areas, and is most apparent in the larger towns, particularly Port Moresby, where urban social networks have most obviously broken down. Growing unemployment, the rise of the informal sector, including begging and crime, and a range of anecdotal evidence suggest that urban poverty is increasing.

Whilst the greatest extremes of poverty are in settlements, not all those in settlements are poor. In Blacksands, most households had incomes below the national average, and at least a quarter had problems meeting school fees, paying rents and providing food; some were indebted to kin or friends and some children missed school. Almost half the households supplemented cash incomes with subsistence food gardens: as one woman said: 'If my family didn't have a garden, we would not be able to live in Vila. It is very hard to get a job . . . most of the time we have to depend on the garden and on the sea for food' (quoted in Mecartney 2001: 156). Having a garden, as in Port Moresby (Battaglia 1986) makes their owners feel productive and enables greater self-esteem. Agriculture is practised in every urban area, from labour-intensive cultivation of root crops and flowers in 44-gallon drums in Funafuti, to extensive and diverse production and marketing in Suva, where about 15 per cent of the urban area is under cultivation. There food gardens occupy land next to homes, distant idle land supports less intensive use and there is adventitious cultivation of available roadside verges and parkland. A significant amount of food comes from such gardens, invaluable where urban poverty is worsening and nutrition deteriorating (Soulsby and Murray 2000; cf. O'Carroll 1997). More would have gardens if access to land were possible. There, as elsewhere, women were vocal about men 'wasting' money on alcohol, cigarettes and *kava*. In Suva, where one out of eight people lived in settlements by the mid-1980s, about 12 per cent of these households were living in 'cash poverty' (with incomes of less than F$63.10 (US$40) per week) at the end of the decade, and as many as 44 per cent two years later (Bryant 1992, 1993). Here, as elsewhere, households with the most diverse income sources (including remittances) fared better than those with only one source. Larger households have tended to fare better, as customary extended household structures have come under increased pressure, and rural 'subsistence safety-nets' have declined in significance. In Vila, as in Suva and Port Moresby, the parallel 'urban safety net' of extended kinship (Monsell-Davis 1993) was also fading fast, as demands on relatives increased; as one woman in Blacksands observed: 'The family on the island and the extended family here in Vila always think they can just come and stay whenever they want. The extended family needs to understand that straight family comes first in town' (quoted in Mecartney 2001: 137). Indeed many urban families supported the idea of sending long-term unemployed youth back to the villages.

In Suva, where Fijian urban unemployment ranged from 16 per cent to 30 per cent in different parts of the urban area by the early 1980s, coping mechanisms involved an intensification of culture and the revitalisation of village forms of social organisation: a moral economy extending beyond the household, in a complex network of reciprocities, where the employed look after the unemployed in an increasingly elaborate network of ceremonial and social obligations: 'a system of social insurance that yields an average welfare, leaving few to starve, and preventing household accumulation by a small number'

(Gounis and Rutz 1986: 86). In that collective ethic lay a degree of long-term frailty. The popular romantic view of an 'urban safety net' of the extended family, or *wantok* (kin) system, is no longer valid (Monsell-Davis 1993; Government of Fiji and UNDP 1997: 89–91). Amongst migrants from Pollap atoll in the tiny town of Weno (Moen) in Chuuk, FSM, new patterns of inequality in access to income and goods have produced 'a combination of public sharing and private hiding' and new tensions in social relationships in a society where generosity and sharing food were highly valued (Flinn 1997: 149). There and elsewhere new access to jobs and education have produced accentuated social and economic differentiations, growing tensions for old forms of reciprocity and lingering, if often inaccurate, notions of egalitarianism.

Whilst many urban households have some access to subsistence incomes and do benefit from kin support, long-term disadvantaged households, who have lost their ability to reciprocate favours, may be disowned and ignored even by close relatives:

> A familiar sight in Papua New Guinea's towns are aging, and sometimes decrepit panhandler and bottle collectors. Less visible are elderly hospital patients with apparently no close relatives to attend to their needs, and older people who have been dumped in town to be looked after – or not – by migrant kinsmen.
>
> (Zimmer 1990: 205)

Growing old in Melanesia no longer confers special privileges. As incomes are increasingly important for food, clothes or school fees, poverty has become of greater socio-economic significance in the absence of adequate or even minimal welfare services. Whilst migrants come to participate in the cash economy, the terms of that participation are so uneven that some people end up struggling to pay for a rented room, and a basic meal of rice and tinned meat or fish: '*Mifala wok long kaikai nomo* (We're just working for food)' (Mitchell 2000: 192). Low incomes and lack of support during sickness or unemployment gave a sense of biding time, waiting on unforeseen and uncertain opportunities, and sometimes securing multiple jobs, maintaining strict budgets and abandoning some 'traditional' obligations, simply to get by. Many residents survive rather than prosper in the city. A number of other situations are indicative of poverty, and outcomes of it. Suicide levels and domestic violence are more common in the larger towns. In PNG, the *rabisman* (a worthless person or beggar) has become more apparent, whilst the cities have become new centres of high incomes, conspicuous consumption and destitution, and sites of growing inequality.

In Fiji in the 1990s around one in every four households was living in poverty, assessed in terms of income levels, housing status and the ability to purchase an adequate diet, and that proportion was almost exactly the same in the urban areas. While the Fiji economy grew by approximately 25 per cent between 1977 and 1991, the proportion of the Fiji population living in poverty actually grew by around two-thirds: minimal 'trickle down' had occurred, even in a period of relative prosperity, and poverty was actually worsening. Disparities were most evident in the urban areas, and more obvious than they had been a decade earlier:

> Against the privileged urban groups who derive their incomes largely from salaries and business activities, there are the urban poor, suffering from unemployment, or deriving low incomes from casual employment, from diverse self-employment activities in an emerging informal sector, or from wage jobs that are not protected by minimum wage legislation.
>
> (Government of Fiji and UNDP 1997: 45)

Three years later it was estimated that a further quarter of the population (probably mainly in rural areas) had slipped below the poverty line because of both droughts and floods (Fiji Council of Social Services 2000). After the 2000 coup the situation worsened again.

Throughout the region traditional social networks for the support of the poor, especially in urban areas, are no longer as strong or comprehensive as they once were. In Fiji there are a growing number of applicants for the Family Support Allowance (only available to the poorest of the poor), an increasing number of female-headed households (and their over-representation amongst the poor) and a growing demand for institutional care for children and elderly people who do not receive adequate family care (Government of Fiji and UNDP 1997: 10). The poorest households are characteristically the smallest, with no more than one income earner at the most, and often isolated from other kin. Poverty, itself a new problem, especially in terms of nutrition and health, contributes to other disadvantages: isolation, a sense of social inferiority, frustration, lack of opportunity, physical weakness, vulnerability and humiliation (ibid.: 66), and these characteristics are increasingly passed between generations. That one of the more vibrant economies of the region exhibits substantial urban poverty problems implies that these exist elsewhere. Only strong linkages between the urban and rural sectors, and the relatively small size of other urban centres, prevent urban problems being as visible as in PNG and Fiji.

Income inequality is often considerable in urban areas, but its significance is difficult to assess where many households have some access to subsistence resources (see, for example, Atoll Research Programme 1996). However, over time, urban households have become more dependent on the cash economy (partly because of the declining availability of urban land and growing pressure on coastal marine resources). In Vanuatu, for example, a high proportion of the incomes of the urban poor consequently goes into food and rent, and a disproportionate share of the tax burden falls on low-income households since they spend a higher proportion of their income on highly taxed imported food (Bryant 1993: 38). Similarly in Majuro, where access to fresh food is expensive and difficult, there is a high incidence of nutrition-related diseases such as kwashiorkor, malaria and diabetes, and 'the average twelve-year-old child living in Majuro and Ebeye has five decayed or missing teeth and there are frequent abscesses and facial swellings' (ibid.: 40). Expenditure surveys indicate that particular consumption patterns may prevent adequate nutrition; thus in urban Vanuatu, husbands 'generally spend up to 50 per cent of their income on alcohol and cigarettes with obvious implications for the families' (ibid.: 47). Inadequate urban diets are a function of poverty, alongside minimal access to land and fresh food, a lack of knowledge of nutrition and a transition away from breast-feeding (Coyne 1984; Connell 1989). In Blacksands malnutrition was a growing problem, as diets became dominated by rice, rather than 'island food' – root crops and green vegetables – and calorie intakes were small, with a low nutrient density (Mecartney 2001). In urban centres, otherwise as different as Majuro and Port Moresby, there is a lack of local produce. By the 1970s some three-quarters of all food consumed in Port Moresby was imported (Flores and Harris 1982: 147) and comparable proportions exist in other towns. In Port Moresby especially, food imports constitute:

> a continuing link between the city and the coloniser long after the formalised connections between the two are broken . . . a by-product of colonialism has been a heavy dependency on Australia for the provision of fresh fruit and vegetables.
>
> (Forbes 1975: 17)

The rapid subsequent growth of Port Moresby emphasised this situation, even as the real price of most imported foods moved consistently upwards, and the range of processed food increased, after independence. There as elsewhere, urban nutritional inequalities are as great as income inequalities.

Whilst formal evidence of urban income inequality, and other measures of anomie – such as poverty, malnutrition and social distress – are rare, in many towns there is a range of visual and anecdotal evidence of difficult urban circumstances, summed up in the case of Fiji:

> More Fijians are squatting and homeless and landless in towns than ever before. More are found in homes for the destitute, the aged and for abandoned children and living on the destitute allowance. There is a widespread misuse of drugs and alcohol throughout the country (not only in urban areas) and the incidence of a variety of crimes is increasing. For example, the number of arrests for drug possession rose from eight persons in 1983 to 199 in 1988.
>
> (Bryant 1993: 56; see also Bryant-Tokalau 1995)

Sober academic accounts pale in comparison with those of itinerant journalists and travel writers, especially for towns like Ebeye, where efforts at urban management are yet to transform the awful urban conditions that existed prior to independence. Effectively the island has now been transformed into no more than a squatter settlement:

> Ten thousand Marshallese lived here on an islet of 40 hectares. The children were as thick as flies, the plywood barracks so close you had to sidle between them to get to the shore . . . the density of garbage gave it the appearance of a clumsy land reclamation project.
>
> (Evans 1992: 262–3)

Whilst Ebeye's urbanisation problems are probably the most difficult in the region, few Pacific states have managed to avoid stress as urban problems, on a smaller scale, come closer to those in other developing nations.

Access to formal sector employment, even in the public sector, is substantially influenced by kinship ties, particularly in Melanesia. Most urban poor and unemployed have some support from kin, particularly where it is most needed, in the larger towns. Otherwise associations created and sustained by ordinary people to protect and promote specific interests, including trade unions, are largely absent. The close rural ties that are maintained by many long-term urban residents are not merely of sentimental value but stem from the intention of migrants to return, as a response to the bleak prospects of access to housing, insecurity of urban land tenure, low or no pension payments and health benefits after retirement from wage employment. They also relate to notions of obligation and identity. One result of the maintained strength of ethnicity is that the urban poor are not yet in a situation where they can cooperate to defend their social and economic interests.

Sustainable economies?

The narrow yet open economies of the region are increasingly influenced by transitions in the international economic system, and particularly the superior growth of Pacific rim nations. Island states have established stronger economic and political ties with Asian states since the 1980s. Foreign investment, tourism, new communications and intensified trade have drawn the Pacific islands more comprehensively into the global system. Increasingly free trade has seen the gradual dismantling of preferential trade agreements. The challenge to the region's economies remains the need to create employment for rising populations with higher expectations, cope with international fluctuations in demand, trade and economic growth, restructure and diversify domestic economies, and gain greater international competitiveness. Declining aid levels emphasise the need for restructuring and privatisation (but in circumstances where private sector profitability is limited) though this reduces public sector employment. Several states were facing economic crises in the mid-1990s; in particular cases these contributed to or engendered serious political crises.

As rural development generates less economic growth and aspirations shift, urbanisation has become more evident within the Pacific (and, through emigration, in metropolitan Pacific rim states). The island states have sought greater economic diversity through the growth of manufacturing sectors, but without success. Industrial growth is likely to be even slower in the future, despite the liberal terms offered to overseas investors, with the decline of concessional trading schemes. Diversification will be more difficult. Tourism offers prospects for development, but inadequate transport links and intervening opportunities have restricted possibilities.

Rapid population increase, especially in Melanesia, the limited capacity of the formal sector to absorb the projected growth of the labour supply, stronger aspirations towards formal sector employment and deep-seated impediments to raising productivity in traditional activities, often associated with land constraints, continue to restrict economic growth and development. Employment growth rates are likely to continue to be outstripped by population growth rates. Even with economic growth the burden of absorbing the majority of the labour force has fallen on the subsistence and informal sectors, neither of which receives significant support from any island governments. Increased unemployment and a growing incidence of poverty have been the outcomes.

Uneven development is unlikely to decrease, but whilst poverty is evident in several states, few governments recognise its existence. This has posed substantial problems of deprivation and inadequate nutrition and influenced increasing urban crime rates, a far cry from when a book on Fiji, Tonga and Samoa could be entitled *Where the Poor are Happy* (Owen 1955). The rhetoric of self-reliance, at national and household levels, has disguised a situation in which there has been a growing dependence on external sources of funding, whether from aid, remittances or investment. This has, in part, contributed to new forms of socio-economic inequality in cities, and incipient class formation, though ethnic and regional divisions and traditional power structures are of pervasive importance. Informal employment is growing faster than formal employment, as it must, resulting in more complex urban economies – and more complex urban planning and management issues for public services, which are no longer growing. The combination of weak economies, overburdened bureaucracies, urban unemployment, fractured social networks and uneven development challenges notions of sustainable development.

Part 2
Urban issues

5 Managing urbanisation

> I think one of the reasons why there are 12 town planning reports gathering dust in our archives, is that under our present land tenure system it is perfectly possible for a very wealthy household to build a mansion next door to a very poor household. . . . If you zone towns with the best will in the world, and a government made up of angels, you are still going to end up with better services to those areas which are wealthy than you would provide to those areas which are not. With this rather impossible mixture we have now, we are more or less forced within our limited ability and resources to provide the same services for everybody, and I'd like someone to come along with a town plan which does not destroy that. I don't think we are ever going to have a town plan ([laugh]).
>
> (Crown Prince Tupouto'a of Tonga, reported in *Matangi Tonga*, March–April 1993: 11)

Planning in industrialised societies is an integral part of the development process and viewed as an essential management tool. It should precede implementation of most aspects of government policy. In democratic systems its legitimacy and acceptance rest on considerable public participation in the planning process. But urban planning in the Pacific is mostly planning without teeth: official guidance that includes standard features of development control like the spatial zoning of land use, the adoption of modern building codes, the reservation of public open space, and even the inclusion of measures directed at environmental responsibility, but with wholly inadequate implementation. Many of the trappings of western municipal planning have been transferred to the Pacific, at least in the sense they are apparent in formal plans and procedures, but the crucial proscriptive ability of urban administrations to prevent undesirable development, backed by legislation, is largely absent. Even the best plans are useless when they cannot be implemented. Exceptions exist, where some successes in managing the course of urban development have been achieved, in the largest cities: Suva, the National Capital District (Port Moresby) and, recently, in the atoll environment of South Tarawa. Various reasons account for the greater persistence of failure, many related to the political composition and will of governments in small, relatively poor and relatively traditional societies where little is done to weaken or compromise the influence of local landowners in and around the towns. Ironically, aid agencies and donor countries continue to promote the need for better plans and better legislation when even those Pacific states that already possess both find it so difficult to put them into practice.

In 1982 a report on urban management in Kiribati was commissioned by the United Nations to investigate the emerging crisis in South Tarawa (Larmour 1982: 2–3). The six main issues that emerged were similar to those found in almost all Pacific urban areas and their intractable nature has made them equally relevant today: the relationship between national and local government in the national capital; the problems and opportunities that would follow when separate local government areas on Tarawa were linked (by causeways); the local governments' revenue base; the possibilities for contracting out services (such as refuse collection) to the private sector; the role of the land tenure system; and ineffective physical planning under the Land Planning Act.

Urban planning in the Pacific has passed through several identifiable stages depending on the degree of urbanisation of particular countries. Producing plans is usually left to foreign experts, as is the drafting of legislation to implement them in the few (mainly Melanesian) countries that have adopted such measures. Here as elsewhere major initiatives by the World Bank, Asian Development Bank and other agencies to promote better urban management have multiplied (see, for example, World Bank 1995a, 1995b, 2000a). Such initiatives include the ADB-supported urban management strategy studies and plans for Port Vila (Vanuatu) and South Tarawa (Kiribati). The situation in Port Vila is indicative of the pressures increasingly directed at achieving more adequate urban planning and management. Few planning permits had ever been issued by the city council, there was little enforcement of existing requirements associated with land use and no real attempt or capability to plan for the growing informal settlements on the urban periphery. After years of national neglect of urban affairs and the lack of an effective strategy to handle urban growth, tentative movement towards a comprehensive urban management strategy brought a management implementation plan through which effective planning procedures could be introduced, together with a legislative framework to ensure implementation and enforcement, prepared in 1997 by a Hong Kong-based consultancy group in conjunction with the Vanuatu National Planning Office. Two workshops were held in the city, each of which was well attended by government, traditional leaders of urban villages, the private sector and NGOs, and considerable publicity was given through the local media. Though a 150-page Urban Management Report appeared in the following year (Asian Development Bank 1998) there is no indication that any part of it was implemented in policy or practice. The government was focused elsewhere. In effect, despite considerable expense and a flurry of externally inspired activity, every government agency and the municipal authority responsible for planning in Port Vila were almost exclusively confined to processing planning applications, had little or no resources (human or financial) for developing and reviewing a plan or had other

Plate 5.1 Port Vila, Vanuatu. © Philippe Metois

priorities. All 'planning' remained reactive rather than proactive. Here, as in most other Pacific towns, despite critical urban problems, planning was largely instituted from outside the country and was not implemented within it.

Common problems face urban management and planning in the Pacific, though each country exhibits distinctly different characteristics based on divergent colonial experiences and more recent geographical, demographic and economic factors. All the Melanesian countries, for example, possess land use planning legislation and the larger towns are managed by local government councils based on British municipal models. In Polynesia the lack of conventional planning legislation or urban municipal government has distinguished Polynesian capitals from their counterparts in Melanesia. The need for controls and guidance over development, together with the task of creating a suitable urban institutional framework to carry them out, has been widely recognised as essential for resolving urban development problems in Polynesia (Ratcliffe and Dillon 1982; Western Samoa Department of Public Works 1984). Deficient urban administration, like that of Port Vila, also exists in the state capitals of much of Micronesia, but here, apart from Kiribati, there is an American legislative legacy in the character and features of municipal government and a different fiscal situation brought about through the funding provisions of the Compacts of Free Association with the United States. Such differences are evident in the privatisation of various utilities in the Micronesian towns.

Although urban planning and zoning controls are introduced to enable orderly development and distributional equity for such municipal functions as service delivery, most western cities where such controls exist are characterised by substantial spatial inequality in the provision of social and physical infrastructure. Pacific towns and Polynesian towns in particular are, by comparison, collections of urban villages and tend to lack the more well-defined spatial income distinctions found in other urban centres. Though Pacific towns may demonstrate a degree of social and political equity, hence the arguments of the Crown Prince of Tonga quoted at the beginning of this chapter, they also include areas of severe economic inequality represented by the Sopu swamp dwellers of Nuku'alofa and many of the informal settlements in Port Moresby and elsewhere. The extent of such marginal and marginalised areas is increasing.

Contemporary Micronesian towns are largely unplanned, reflecting different colonial influences, and have weak institutional development control processes. Indeed, there is very little evidence of urban land use planning in Micronesia apart from the British administrative legacy seen in the South Tarawa Urban Management Plan in Kiribati (Jones 1995), and a land use zoning plan for Pohnpei (Pohnpei State Land Use and Zoning Master Plan Taskforce 1995). The only physical examples of planned urban development in the region, other than mining towns, are the tiny, yet grandiose capital of FSM at Palikir in Pohnpei and the proposed new capital for Palau. Ironically, this could result in a situation where two of the smallest Pacific states possess the most formally planned urban centres in the region. Despite such grand schemes, comprehensive urban land use planning is otherwise absent. The establishment of more effective urban governance and planning is challenged by land tenure realities. This in turn closely relates to the basis of traditional power and the way in which it is exercised in Pacific countries: the adoption of land use planning legislation is often seen locally as a means of weakening control over land held by traditional authority and placing more powers in the hands of government and the bureaucracy. Although both governments and the bureaucracies are also dominated by major landowners, this does not make land transfers any easier – indeed it makes them much more difficult – resulting in something of a paralysis in local government.

Attempts to introduce controls over development in newly established outposts date back to the earliest stages of colonial settlement. Among the first examples of urban planning in the Pacific was Apia, brought under a form of development control by the Municipal Convention of 1879, whereby Germany, the United States and Britain negotiated the establishment of a self-governing urban enclave. Control of foreshore land of the kind found in the Apia municipal regulations was innovative at the time, scarcely existing even in Britain and its Australasian colonies, and rested on enforcement where necessary in a magistrate's court:

> Since it is desirable for the public health that the sea breeze should circulate freely through the town and not be intercepted by buildings near the

> water along the shore of the harbour, all persons are hereby warned against erecting or repairing buildings or structures of any kind whatsoever on the north or sea side of the present public road. . . . Any violation of this Regulation shall be punished by a fine not exceeding $100 or three months imprisonment with or without hard labour.
>
> (Apia, Western Samoa, *Municipal Regulations*, 30 January 1880, cited in Pringle 1989: app. 1)

More than a century later Apia possesses a wide range of conventional urban development regulations but has no municipality or municipal government, no enforceable town plan and weak building controls. Many of the features of western urban regulation are in place but the planning and development process, once evident in the colonial period, has long since broken down amidst widespread local concern, as an editorial in the national newspaper commented:

> Whoever got this thing [new government office building in Apia] to be built on the waterfront anyway? All it's doing is block out the view thus shortening everyone's vision. These games power-hungry men play are maddening, maddening . . . and it has always been the case, it's the ordinary people who end up carrying the burden. This is something everyone should think seriously about.
>
> (*Samoa Observer*, 4 March 1992)

Management structures, policies and procedures under such conditions are no longer evident and therefore functional.

In time, town planning legislation modelled on British (1932) legislation was introduced to many former colonies 'as a convenient response to local political pressure for better living conditions, at a time of international crisis when the British government needed to ensure the loyalty of colonies for the impending Second World War' (Home 1993: 407). This less than altruistic introduction of urban planning saw it emerge in the post-war world order as a palliative used by some colonial regimes to promote better urban living conditions and delay nationalist pressures for decolonisation. Urban planning actively constituted a form of power derived from professional and technical knowledge and exercised by colonial and later post-colonial regimes and donor agencies (Mitchell 1988). As far as the colonies were concerned, the brand of urban planning promoted during the independence era of the 1960s and 1970s:

> proved to be an inadequate technical response to massive social and political pressure for change, and indeed was recognised as such by those planners and colonial officials who formulated it at the time – a policy of expediency. Not only was town planning inadequately resourced to deliver real improvements on the ground, but the relationship of physical planning to wider issues of development planning or social welfare was confused in the colonies.
>
> (Home 1993: 407)

This confusion has never been dispelled in the Pacific and has complicated and delayed efforts by external agencies to implement standard forms of planning and legislative procedures in all the regional capitals.

In this chapter the experiences of urban management and planning are illustrated with particular emphasis on the local resolution of contemporary issues. First, the financial issues confronting urban managers, illustrated by the situation in PNG and the plight of provincial towns in particular, enables a focus on the complex issues surrounding the fiscal base of urban planning. Second, difficulties in implementation are illustrated in the still unresolved struggle to establish planning legislation and develop a nexus between parliamentary and customary laws in Samoa. Such issues are very typical of the situation in most smaller countries where a formal tier of urban local government has never been established. Third, an innovative example of urban management planning for the fragile atoll environment of South Tarawa in Kiribati is discussed to illustrate the manner in which some strategies that are sensitive to the objectives of sustainable urban development have been formulated, though not fully implemented.

Financing urban development

A necessary condition for local government to function optimally is that it operates on a firm financial footing. Unfortunately this is virtually non-existent among urban entities in the Pacific, as recent experience of local government in PNG makes plain. Without adequate capital and resources even the best plans and management strategies – and there are few enough of these – cannot succeed. The urban authorities that exist are substantially assisted by central government grants in their responsibility to provide urban infrastructure and services because municipal rates are low and difficult to collect. The quality of service provision and delivery has suffered as a result, with various forms of corporatisation or privatisation of public services being tried in order to improve the situation. Four main kinds of privatisation have been identified in developing countries worldwide and all have been attempted in the Pacific: divestiture, which is the situation when government assets are sold or transferred to the private sector; contracting out, when government enters into contracts with private operators to supply services, or to manage certain activities; financing, when public sector projects are financed by the private sector; and deregulation, when private enterprises are permitted to supply services that were previously only supplied by the public sector (Lee 1997). In the Pacific these range from the reintroduction of a French privatised water supply system in Port Vila that was in place before independence, to a controversial Build-Own-Operate-Transfer (BOOT) arrangement for urban water supply in Port Moresby (the breakdown of which led, in 1996, to political accusations of incompetence and even fraud).

The operation of local government in PNG, where there is a greater range of towns than elsewhere in the Pacific, reveals some of the problems of urban finance. Funds were primarily allocated by national government to the National Capital Development Commission (NCDC) in Port Moresby and the former provincial governments, though there was also some direct expenditure through line departments (most obviously housing) to existing town authorities. Outside the National Capital District (NCD) the main branch of government responsible for urban finance has hitherto been the provincial governments. This presented several problems: first, there was substantial mismanagement and maladministration in these governments to the extent that most were suspended at one time or another (which provided the central government with a reason for their demise in 1996); second, although provincial governments collected revenue from charges such as land rates and vehicle taxes in urban areas, this was not necessarily spent there. Sometimes this neglect of the towns stemmed from perceptions that funds would be 'wasted' on migrants from other provinces, or would improve urban infrastructure and thus stimulate large increases in rural–urban migration. Third, actual revenues of town authorities were significantly less than expected because of poor collection performances and difficulties in dealing with provincial governments (Asian Development Bank 1992: 29). Urban authorities were consequently seriously underfunded for the tasks they were expected to perform.

A substantial proportion of urban income in PNG is generated locally, which in the case of the NCDC came mainly from land taxes, retail sales taxes and water and sewerage fees, with some 85 per cent of total revenue being raised from taxes and fees within the city itself. By contrast in the second largest city of Lae just over half of its much smaller urban income was raised locally, and the problems that emerged there were typical of all towns in PNG outside the capital. Financial sources previously available under the Local Government Act of 1963 were acquired by the provincial government without any reduction in the council's responsibilities:

> There is no established proportion of provincial income generated in the city transferred to the Lae City Authority, as the provincial government is consciously pursuing a policy of using urban-raised revenue as a means of financing rural development. Other taxes which had been originated by the city of Lae, a tyre tax and a fuel tax, are now also collected by the provincial government.
>
> (Axline 1989: 32–3)

The decline in funding had 'led to a deterioration in urban infrastructure and services'. In 1991 as much as K7.5 million (US$7 million) was generated within Lae from charges such as liquor licences and vehicle registration but spent elsewhere in the province. Medium-size towns like Goroka and Rabaul have failed to become self-financing in spite of moves towards user-pays charges and improved financial management. Smaller urban centres have even less ability to generate local finance because economies of scale in terms of service provision or cost recovery decline sharply.

The outcome of these urban financing difficulties is that necessary management is often inadequate. In Lae, where the provincial government allocated relatively little money to the urban area, few management functions were undertaken and service provision was limited. There was some connection between the lack of urban services and tensions between different ethnic groups within the city that culminated in serious urban violence in October 1992, when fighting between groups of settlers and the longer-established Morobe population resulted in several injuries and the destruction of many homes. The national newspaper fulminated: 'Since independence, Lae has gone from a booming industrial and commercial city to a rundown outback town ruled by a minority with little respect for the rule of law and the welfare of the community as a whole' (*PC*, 6 October 1992).

Similar if less extreme situations occur in towns throughout Melanesia. The Mount Hagen Town Authority reported that in its first full year of operation (1990):

> it could not implement any project because all projects submitted in the budget were not funded by the Provincial Government. Further the small grants received from both the National (K300,000) and Provincial (K200,000) Governments were insufficient for any new projects. The only project which was undertaken was the complete upgrading of the Public Toilet near Hagen Coffee Factory at a cost of about K8,000.
>
> (Mount Hagen Town Authority 1991: 9)

In this case, and in others such as the Home Finance Corporation of Solomon Islands, revenue has been almost wholly absorbed in maintaining the institution itself (in salaries, vehicles, building maintenance, and so on) rather than achieving management goals or implementing specific tasks.

Good urban management is compromised by the manner in which available funds are allocated by national and provincial governments and the various line departments and ministries. Urban management has become fragmented and complicated in a region where bureaucratic processes are already invariably slow and inefficient. Again, a typical example is the case of Mount Hagen:

> The Members of the Town Authority are gravely concerned over the problems associated with development in the town. Of particular concern is the present fragmentation involved in deciding the 'use' of land in the town. Three separate 'Boards' are involved: Land Board, Town Planning Board (Committee) and Building Board, each established under separate Acts of the National Parliament. The major problem is that each respective body does not coordinate with the other, because: there is no common membership on each board, in some respects members lack expertise, there is little or no meaningful dialogue between the respective bodies. Due to these legislative deficiencies, the lack of a town plan for the use of Mount Hagen, and the fragmented system of land use development, and because

> the situation has existed for many years, we are now faced with an insufficient sewerage system, inadequate distribution of water supply, no drainage system, crowded settlements on the fringes of the town.
>
> (Ibid.: 9–10)

Payments collected for services such as sewerage and garbage were generally in arrears and thus generated much less revenue than anticipated and necessary to maintain full urban functions. By mid-1992, little more than three years after it had been established, the Mount Hagen Authority had outstanding debts of K500,000 (US$480,000) in land rates and service fees for sanitation and garbage, with the debtors including businesses, churches and individuals (*TPNG*, 18 June 1992). This was typical of the situation confronting urban authorities in PNG which were without the resources or administrative skills to ensure that debts were paid promptly, or at all.

Despite the ubiquity, visibility and gravity of such urban service provision problems, solutions were usually delayed until crises emerged. Responses have therefore been partial, expensive and unable to address wider problems. Urban road construction and water supply provide the most dramatic examples. In Port Moresby in the early 1990s a government decision to construct the Poreporena freeway at the huge cost of K62 million (US$50 million) was investigated by a parliamentary inquiry. (This paralleled a proposal to construct a similarly expensive four-lane highway through Honiara.) In neither case were traffic loads on existing roads normally excessive. Similarly, persistent water supply crises in Port Moresby during 1992 led the NCDC to seek cabinet approval for a K70 million loan to implement an improved system, and contract out a K2 million 18-month study of management of the water supply system until the year 2012. The upshot was the award of a 22-year concession to a Malaysian company under a BOOT arrangement whereby the company provides the capital, then builds, owns and operates the project until its investment is recovered and profitability achieved; ownership is then transferred back to the municipality. Questions about the operation of the project and financial implications for ratepayers prompted a government review in 1995; serious accusations were made about poor management and a very disadvantageous deal for the citizens of Port Moresby (*Saturday Independent*, 21 October 1995). Questions were also raised over the way low-income consumers might fare under privatisation of water supplies in Port Moresby, and how successfully the NCDC would take responsibility for water provision when the company concession expired.

Urban management in Melanesia has often been crisis management rather than achieving a level of 'good housekeeping' that might avert the worst crises. Grandiose and dramatic solutions to urban problems have outnumbered planned and monitored responses of more modest kinds. Finance for urban infrastructure maintenance is rarely found locally and is unavailable until the crisis occurs. Major external assistance was proposed for PNG in 1992, with ADB agreement to provide financial support for the expansion of much-needed services in six main provincial towns. The money was sought to improve their attractiveness as employment centres and thus reduce migration to the largest towns in PNG. The Bank saw a particular role in institutional strengthening and capacity-building, together with supporting projects that promoted urban management, employment generation and expansion of regional centres. Specifically, the four largest towns were to be developed through local government assistance, preparation of structure plans and identification of priority projects; further help was to be granted to land management through support for the Land Mobilisation Programme; institutional strengthening of the National Housing Commission (NHC); and determination of training requirements (Asian Development Bank 1992: ix). Implementing these proposals has proved almost impossible; the dilemma being that action must be taken to upgrade employment opportunities and living conditions in the towns, but that this may encourage increased migration to Lae, Mount Hagen and Port Moresby in particular. The principal means of dealing with this dilemma has been to ignore it and take no action, thus increasing the likelihood of worsening urban conditions and the eventual need for crisis management, whilst avoiding short-term fiscal demands on financially fragile urban authorities.

Concerns are frequently expressed by governments and NGOs about the extent of intervention in domestic affairs that results from the conditions placed on financial assistance arising through international aid. High levels of aid dependency in the Pacific have ensured extensive interventions at all levels of government with the ADB, IMF, World Bank and bilateral donors having had a major influence on policy formation. Resolving difficulties in land supply in PNG through the introduction of land registration, for example, was a controversial IMF Structural Adjustment Programme condition in 1996. Similarly, the preparation of regional development plans in Polynesia in the mid-1980s came about at the request of bilateral aid donors from Australia, EEC and New Zealand and resulted in the first land use planning studies in Tonga (for the Vav'au and Ha'apai island groups). Very few urban centres in the Pacific are close to paying for or maintaining adequate urban infrastructure and services and are thus characterised by limited autonomy in their conduct of municipal affairs. A key challenge is thus to reduce costs of urban management to what is locally sustainable.

Parliamentary and customary law

Donor governments with strategic interests in the Pacific have been among the first to recommend the institution of western urban planning legislation as a means of regularising and controlling urban development. In Melanesia this objective has generally been achieved by the parliamentary adoption of such laws after independence but major difficulties, already alluded to, have been experienced in their implementation. Elsewhere the situation is different, with towns in the former British-influenced countries in Polynesia and Micronesia having adopted some measures and others, such as those in the former US-administered Micronesian states, being governed under municipal codes that do not include comprehensive planning laws. The 60-year experience of events in Samoa provides some insights into the nature of the debate over urban development that took place in the first island state in the region to gain independence.

A succession of attempts to institute modern urban planning procedures in Apia demonstrates the attention given to this over a long period. The difficulty of establishing a modern system lies not so much in any failure to appreciate that such a change is necessary but in grappling successfully with the underlying factors which have ensured that efforts at reform have met with defeat. Attempts by development funding agencies and aid donors to introduce a planning regime backed by legislation appear never to have investigated why Samoan governments have been so resistant to changes affecting controls over land title. Underestimation of the strength of local social organisation on the part of many external advisers and consultants perpetuated an extraordinary succession of unfulfilled recommendations dating back, in the case of Apia, to the colonial period before the Second World War.

Two decades later, well before independence, an attempt to confront the challenge of physical planning resulted in the establishment of a Town Planning Committee. The terms of reference recognised most of the important planning and development control issues facing the town and the committee was required to place before government 'firm recommendations [and] such estimates of cost, urgency and priority it feels competent to give' (cited in Western Samoa Department of Public Works 1984: 2.3). The committee met just once in 18 months because it was not provided with material to consider. In 1956 it was recognised that urban planning would be fruitless unless a local authority was constituted to administer it, and the first steps were taken to achieve this through an Apia Advisory Committee. The Minister of Lands served as chairman, and committee members included the Directors of Works and Health and the Chief Surveyor. In the words of the advisory Apia town plan of 1984:

> Unfortunately, the Chief Surveyor was not relieved of his other duties, and after an initial start with the recommendations regarding traffic patterns, bus terminals and after raising the question of zoning, departmental preoccupations of his regular job pushed planning further and further into the background. By early 1959 the Apia Advisory Committee had, for all practical purposes, disappeared entirely.
>
> (Ibid.: 2.5)

Even in colonial times, it was impossible to establish an effective urban planning system. The colonial administration was not convinced that tackling the planning issue was of sufficient priority for a small town; it offered few advantages to Apia's more prominent citizens and there were more pressing priorities elsewhere.

Preparation of the first master plan for Apia began in 1964 without legislative backing and was only advisory in its direction of the course of urban development. It was completed by the Director of Works and a town planner appointed to his department in the following year. Thus began the production of a series of physical plans and reports stretching over three decades, none of them backed by either physical planning legislation or the executive powers of local government. Expensive and comprehensive tourism, transportation and urban services advisory plans were also produced, most of which commented on the need for statutory land use planning and the legislative ability to implement it. The limited value of advisory planning was eventually officially recognised by government and in 1972, a decade after independence, Cabinet decided that responsibility for land use planning should rest with the Public Works Department; draft legislation was prepared but deferred indefinitely in 1973 by a Cabinet committee. Another town planner was appointed in 1977 and Cabinet once again constituted a Planning Committee and initiated the preparation of a Town and Regional Planning Act which was completed in 1979 (Western Samoa Department of Public Works 1984). Once again the draft legislation never became law. More than a decade later, in the mid-1990s, the cycle began again with the preparation of yet more draft legislation, designed this time to set up a Town and Regional Planning Board and Regional Planning Unit, which has yet to be implemented.

The reconciliation of traditional land tenure practices with the development needs of a growing urban area like Apia underlies the task of improving land use planning. About a quarter of Apia's urban land area is held under customary tenure (60 per cent is freehold) and almost all of it used for residential purposes. As government land and freehold stocks are exhausted, pressure has increased to develop the customary areas, whilst trends towards nuclear family formation have led increasingly to a *de facto* 'ownership' of customary land, as in several other parts of the Pacific, encouraging governments to come up with new tenure arrangements which would allow some of this land to be used for planned urban development. As one official report noted, the solution must somehow incorporate traditional procedures as well as allowing modern development to take place:

> There appears to be a demand for freehold and de-facto 'ownership' of customary land by increasingly nuclear families. At the same time, there is a fierce desire to maintain Samoa's customary land tenure system as the basis of the *fa'a Samoa* [tradition]. Failure to address this issue and to provide access to land in a way which does not alter the basic principles of customary tenure is likely to lead to increasing land disputes and potential reduction of access to land for those with least resources who have traditionally been guaranteed land for their housing and productive needs.
>
> (EDU International . . . 1992a: 14)

At the heart of changes aimed at introducing urban planning is the question of land tenure.

These urban management difficulties in Apia are one facet of a broader requirement to achieve reconciliation in the operation of land use controls enacted by parliament and the traditional procedures of *fa'a Samoa*. This need is underlined by the increasing pressure placed on all levels of society in the Pacific to examine and improve environmental practices. In Samoa, for example, customary decision-making effectively controls development over some 80 per cent of the country including the coastline, lagoons and reefs. Two important pieces of legislation, the Lands, Surveys and Environment Act (1989) and Village Fono Act (1990), attempted to codify decision-making in a range of situations affecting the environment and natural resources, and went some way towards recognising the realities of village-level decision-making and bringing together parliamentary and customary legal codes. The Lands, Surveys and Environment Act allows the Minister 'to enter into agreement with owners and occupiers of customary lands for the purpose of protecting their natural resources and environments' but 'does not exclude other ministers and their departments from implementing conflicting objectives' (Cornforth 1992: 1). Moreover, such an agreement with a minister has little force over the actions of individuals if it is not also bound by other laws and regulations.

The effectiveness of environmental and resource protection built into the two Acts is weakened since 'the boundaries which delimit the influence of government, the village, and private land owners do not coincide with environmental boundaries, and there exists neither overall planning law, nor any satisfactory integrating mechanism which might bring the two codes together to the satisfaction of all parties' (ibid.: 2). The same situation applies to the geographical extent of urban services delivery and infrastructure provision responsibilities which cut across the traditional boundaries of villages incorporated in the Apia urban area. This inability to accommodate community-wide planning needs has created a form of bureaucratic paralysis where key problems are recognised and the means to deal with them identified, but good intentions are overwhelmed by difficulties in implementation and control that stem from different regulatory regimes in different places. This problem arises not from a refusal to accept control over personal actions in Samoan society, where sanctions of a traditional kind can be very effective at the village level, but out of an inability to take collective responsibility for pressing development outcomes and needs in the urban domain because this area has no status or function in Samoan society. The *matai* (chiefly) system remains of enormous significance and many people in Apia are also traditional landowners with full authority over land use and, more importantly, legislative control over a range of issues. Thus within Apia there are quite different tenurial and legal systems and attempts to impose contemporary urban management procedures in such a context is fraught with risk and uncertainty without extraordinary levels of consultation. In 1993 the Department of Lands, Surveys and Environment (DLSE) recommended amendments to the Lands, Surveys and Environment Act (1989) which would allow the incorporation of local aspirations in village resource management plans prepared after full and popular consultation.

Such measures, appropriate in a rural setting, provide inadequate guidance for everyday control over development in a large urban area. The task of enacting more interventionist urban planning legislation, which accommodates the urban villages in Apia and is also capable of gaining widespread community support, remains as considerable as it has always been.

Urban governance in Polynesia

In recent times there have been calls to establish a town council in Apia as a means of implementing yet-to-be-enacted planning legislation. This followed increasing frustration with inadequate urban management and the lack of an accountable tier of local government. The debate has been limited but has reflected the emergence of popular demands for action. Strengthening administrative capabilities at the urban level is a region-wide objective in the Pacific but there is no single model that satisfactorily covers the varied circumstances of even the largest towns. In Polynesia the solution may not lie in the adoption of a conventional municipal authority of the kind already established in neighbouring Melanesia. Expressions of community intent are usually a primary precondition for establishing urban local government in democracies (Power *et al.* 1981) and are required before the acceptance of an effective planning system. Two Samoan examples illustrate the nature of local opinion. The first of these called for the introduction of an Apia Town Council as a means of improved urban environmental management:

> But seriously consider how to get rid of the ugliness in town and keep it away permanently. Perhaps a Town Council, as some concerned people have suggested, is the answer to the problem. Such a body could be established independently of the government, but allocated funding under the government's annual budget. As the Town Council grows and becomes more confident of itself, it should then be able to embark on money-earning projects to assist it financially. Companies could be asked to fund projects aimed at lifting the place's image up, or provide materials to enable these projects to be carried out. Also, such rates as motor vehicle fees and driver licence costs paid by residents of Apia and its suburbs could be directed to the Town Council. The Town Council could organise work groups to assist in drainage work, sidewalk resurfacing, lawn mowing, gardening, trash collecting and so forth, drawing on the community's unlimited volunteer work force. This would reduce loitering around town by many young people who will now find life more meaningful, and feeling useful at the same time for being able to make a contribution to the community.
>
> (*Samoa Observer*, 15 March 1991)

Admirable altruism is apparent, but a more probable example of the model that could occur in Polynesia is that in Honiara (Solomon Islands) and Port Vila (Vanuatu). Both are underfunded and underequipped to fulfil the normal range of municipal functions and are impoverished by central governments that have no wish to build up political competition from an effective (and independent) urban administration in the national capital (Connell and Lea 1993). The effects of this situation on urban management present an uninviting prospect for those Pacific countries yet to adopt urban local government.

A year later, an outraged 'Apia slum dweller' put the case again:

> Your editorial of last Wednesday (11 March 1992, *Fugalei swamps and dead baby*) was thought provoking and heart rendering in the thought of an innocent human baby floating in the filthy swamps and being eaten by pigs. We pride ourselves as the paradise of the South Pacific and a God fearing country that attends church twice a Sunday all in white as snow clothing and look what we have done to the babies – eaten by pigs – bahhh!!! I say 'we' because this is a crime created by society and a government which does not care about the welfare of its people, a government which turns a blind eye to this eye-sore slum right in the middle of Apia city. If the government had cared about the welfare of these people who live in these slums, it would have passed legislation years ago on the standards of dwellings that should be built; improved drainage; created an Apia City Council which should draw its representatives from the Apia City Community. . . . Once we have an Apia City Council, then we have a body which should be empowered to visit dwellings within their own small village suggesting improvements and instigating fund raising projects with government support which should provide at least 50 per cent in a tala-for-tala system of incentive to help these slum dwellers.
>
> (*Samoa Observer*, 18 March 1992)

This particular correspondent was also insistent in calling for an elected town council which could overcome government by appointment and particularism: 'That is why many government instigated projects fail, because it does not use the very people whose lives it is meddling around with.' There is thus some evidence in Samoa, at least, of popular support for a more effective means of determining and implementing urban policy. None the less, two years later representatives of community organisations in Apia 'decisively rejected the need for a municipal authority for Apia, asserting that existing community-based structures . . . and agencies could serve more effectively and at less cost' (World Bank 1995a: 80).

A World Bank mission to Samoa in 1993–4 examined the management of urban environmental sanitation services and outlined four possible outcomes if a municipal administration were to be created: first, the possible duplication of services already provided by government utilities; second, an overburdening of the very limited means of providing infrastructure and delivering services; third, the imposition of municipal taxes in a country where taxpayers were in opposition to a newly imposed general value-added sales tax; and fourth, the creation of a local government structure which would not incorporate sufficient participation by the strong *matai* system of traditional authority (World Bank 1995a). Instead, the Bank questioned the ability of the country to acquire the technical staff to run a new municipality, in the context of a brain- and skill-drain, and suggested the priority should be for the Department of Lands, Surveys and the Environment (DLSE) to: 'develop basic ground rules on land use; improve coordination between central government departments and agencies that provide city-level infrastructure and services; and a streamlined process for obtaining building permits' (ibid.: 101). Despite the need for such changes, these actions would merely have modified the symptoms of a much deeper problem. Urban planning in a post-colonial environment was no more able to surmount technical improvements to address wider issues of social and economic change than in the colonial era (Home 1993). During a 1973 parliamentary debate a Member of Parliament had stated:

> The most precious treasure of a Samoan is his land, and if the country is so conscious of its land, I do not blame them, because we now see that lands have been taken away gradually for purposes of the Government for public purposes. I do not intend to support the [Town Planning] Bill to the extent that the people is [*sic*] denied the right to control their own lands.
>
> (Quoted in Western Samoa Department of Public Works 1984: 6.1)

Land tenure placed a clear limit to the development of urban planning. In contrast, the World Bank's approach was technocratic; it argued that the most appropriate strategy to improve urban management would aim to set clear goals, minimise overhead costs and demonstrate accountability. The solution of strengthening the Environment and Conservation Division of the DLSE was merely a 'back-door' approach to coordinated urban management, reflecting the continued need to create a development control system backed by appropriate legislative powers that would command both central government and community support, but without a new institutional context.

Thus the key question in Apia (as in most of the Pacific) has long become how best to implement and sustain urban development strategies, rather than establish city-level government, despite some popular support. Various examples exist in Melanesia of urban councils whose composition and resources make them unable or unwilling to tackle obvious difficulties, such as the regulation of traditional housing areas, or to attract the kinds of revenue needed to accomplish their existing functions effectively. Thus:

> Although Honiara has a Town and Country Planning Board, responsible for physical planning, there are no guidelines on which to base zoning of land and to allocate services, other than the now-outdated, unofficial 1988–1992 Plan. The Board has the power to enact a local planning scheme to ensure orderly urban development. However, it must be developed through fifteen legally-required steps before it can become operational.
>
> (World Bank 1995b: 55)

Urban planning and management remain in their infancy, even in the larger towns.

Urban management in an atoll state

At a very different scale several small atoll states in the Pacific face urban environmental crises where there is little or no room for expansion and dwelling densities are among the highest in the world. Those most affected are Funafuti, South Tarawa, and especially Majuro and Ebeye. The need for effective urban management is obviously crucial under atoll conditions and there is evidence from Kiribati, at least, that suitable planning arrangements can be made, even where national skills, expertise and capital are particularly limited.

The growth of South Tarawa by the 1990s to a population of more than 30,000 spread over 20 km from Betio to Bonriki (Figure 5.1) heightened concerns about the environmental sustainability of urban development in an atoll location. Plans to improve urban management in Kiribati followed the lead taken by the Urban/Land Planning Unit of the Ministry of Home Affairs and Rural Development (KMHARD) to facilitate a joint approach to the problem and greatly increase the effectiveness of the Central Land Planning Board (CLPB). A Land Planning Act had been adopted by the government in 1977

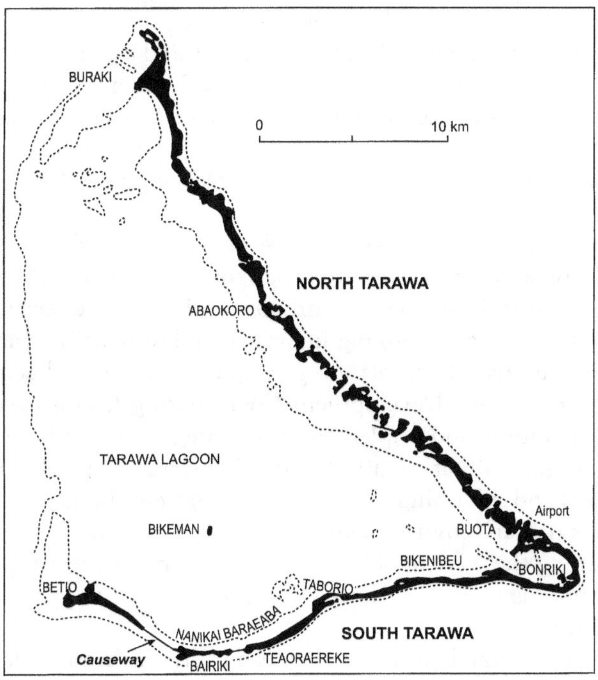

Figure 5.1 South Tarawa, Kiribati

which established the CLPB as responsible for the preparation of strategic plans for South Tarawa. Physical planning was to be broadened in scope through the preparation of a structure plan for the whole atoll, together with altered government representation on the CLPB and more public participation. It was to be more than a decade, however, before serious attempts were made to prepare an urban management plan. The process which culminated in the production of the *Draft Urban Management Plan for South Tarawa* began in 1993 (Kiribati Ministry of Home Affairs and Rural Development 1995). Besides covering the usual sectoral urban development issues, it was the only plan in the region to consider thoroughly the institutional implications of improved urban management:

> The focus of the CLPB is only within the designated village areas – Betio, Bairiki, Nanikae and Bikenibeu – as the remainder of South Tarawa has not been designated by the Government for urban planning and management purposes. As such, there exists no overall or regional management approach for land-use and development in South Tarawa. The CLPB and BLLPB [Betio Local Land Planning Board] have no budget resources for implementation and as planning committees, they ultimately depend on other agencies such as the [Public Utilities Board] to undertake urban development and urban infrastructure works. Given a combination of an absence of high level representation on the CLPB, the lack of coordination and integration between agencies, the absence of a single stand-alone policy document for South Tarawa for use by all key players, and a general lack of commitment to tackling the enormity of the urban issues at hand, the result is that many of the Central Government agencies . . . act in isolation from other players.
>
> (Kiribati Ministry of Home Affairs and Rural Development 1995: 28)

Among the central management recommendations in the *Draft Urban Management Plan* (whose acronym gained little support) were: strengthening local government to enable it to manage urban development effectively; challenging the lack of coordination among interested and responsible parties that had caused the main obstacle to planning, implementation and action; making use of the government's Development Coordinating Committee to obtain a better mix of planning and implementation; integrating land use and environmental planning; designating all of South Tarawa's urban planning needs; reviewing the Land Planning Act to change its emphasis on land management towards urban environmental management, planning and assessment; and obtaining the services of key personnel to implement the plan. Although these were not easily dealt with they provided a realistic appraisal of familiar administrative weaknesses.

National governments are rarely aware of the need to strengthen municipal capabilities to undertake land use planning functions effectively. Most are aware of the inefficiencies of a poorly developed market in land but are yet to address the implications of improving it, such as dealing with the compensation claims of traditional owners and completing land title registration (Chapter 6). The government of Kiribati commissioned a study of institutional strengthening of the Lands and Survey Division in KMHARD in 1996 to assist it in fulfilling its urban planning functions (Jones 1996), hence the problems attached to institutional strengthening have been more comprehensively identified here than elsewhere in the region. Managing the growth of South Tarawa demands the ability to coordinate cross-sector urban planning issues and especially the efficient provision of new water and sewerage projects. In turn this involves dealing with several practical issues including cross-ministry coordination; attracting funding to provide required staff and expertise; and employing a multidisciplinary project team that focuses on outputs with more efficiency and accountability. In particular there is a need to introduce a land registration system, review leased government land in South Tarawa, and achieve overall public sector reform (ibid.). The very length of this list of necessary precursors to institutional strengthening is indicative of the substantial problems that confront all small island states in their attempts to develop and improve urban management. Yet again issues have been clearly identified; however, as elsewhere in Oceania, translating them from the drawing-board and into the government system is exceptionally difficult.

Underlying tensions between traditional and modern ways are crystallised in urban environments and, although often recognised as a particular problem in the Pacific, such issues are rarely addressed in government policy. Little evidence of popular, or indeed any, local participation in the preparation of urban plans exists, but most planning recommendations at least include holding workshops of some kind with townspeople. Reluctance on the part of most urban residents to be involved in any but the most parochial urban issues is a constraint on popular involvement in the planning process. Consequently, elsewhere in Micronesia, the inclusion in Pohnpei State's Land Use Planning and Zoning Act of 1994 of provisions to encourage local participation through special efforts to improve public education appears unique. Plans were made in 1995 to assemble an education committee to meet community leaders and 'determine the best methodology for conducting a public education and feedback process within each municipality' (Pohnpei State Land Use and Zoning Master Plan Task Force 1995: 1–5).

The five years allotted to complete the island-wide consultations seem excessive by western standards but reflects the need to allay fears held by traditional authorities and to involve local communities. Given the extent of recommendations emerging from most consultancy reports, well evident in South Tarawa, and the antipathy to change on the part of conservative landowners, who include local political leaders, traditional perspectives have tended to triumph over external directives, however altruistic or apparently appropriate.

Take-it-or-leave-it urban planning

Whilst the use of examples drawn primarily from Kiribati, PNG and Samoa is not fully representative of the urban experience in the island Pacific, they point to a considerable degree of commonality. Achieving an effective urban planning system, for example, depends upon the economic circumstances of small countries and small towns that have been deeply influenced by a history of colonial dependency, foreign aid support, migrant remittances and the adopted technology and culture of a distant industrialised world. In some respects urban changes have been dramatic: 'Along those same town roads where fifty or sixty years ago Japanese vendors hawked noodles and rice cakes and cheap alcohol, there are now dive shops and video-rental stores' (Hezel 1995: 366). Managing the outcomes of such changes in an orderly way, and coping with many other pressures, is a huge task and requires identification of the most urgent priorities. Better planning can contribute positively to improving the quality of urban life but cannot provide jobs, halt crime or generate the financial opportunities necessary to pay for sophisticated social and environmental services. Towns which cannot pay for even the maintenance of basic infrastructure are in no position to improve their facilities, provide services or adopt better planning procedures. Basic issues thus include: how to convince national politicians that towns, particularly the capital cities, need to raise more revenue; how to accomplish urban land title registration without unduly alienating traditional authority; how to encourage more public participation and self-reliance in urban affairs; how to raise the technical skills and effectiveness of local government employees and the awareness of councillors (where they exist); how to prevent excessive environmental pollution; and how to foster the emergence of locally appropriate ways of enforcing urban regulations of many kinds. The distinctive nature of Pacific urbanisation influences more specific directions, and there are at least six key common features that have influenced the present specific form of development.

First, Pacific urbanisation is significantly affected by a foreign past dating back over a hundred years whose influence extends beyond the physical form of buildings and neighbourhoods to include the importance of adopted regulations and procedures. Indeed, the importance of such 'unseen' legacies, resulting in changes to custom and economy in former colonies in many parts of the world, has caused some of the most intractable obstacles for post-colonial populations to overcome (King 1990). Among the least visible were changes to local legal and justice systems, new educational and administrative structures, health regulations, and trading laws prejudicial to the existence of informal economies. Many led to increased population mobility, disputes over land ownership, and emerging pockets of urban poverty. Collectively, over many parts of the region the legacy of administrative structures does not work well in Pacific settings. Complicating this experience have been the differential effects on urbanisation of the size and nature of former Pacific island colonies and distinctions in the behaviour of the various European, American and Japanese governments involved. The institutions of Pacific urban administration today are primarily artefacts of earlier colonial entities, as they have been in former colonies in other parts of the world.

Second, land tenure in the Pacific has posed seemingly intractable difficulties for governments trying to manage substantial customary land holdings in and around urban areas. The colonial planning and legislative models of development control usually found in the Pacific were devised under very different political and cultural circumstances and have rarely proved successful. In addition to problems of establishing land title, building regulations and construction permits are rarely enforced with certainty, making an orderly development process virtually impossible to implement. Some progress has been made in overcoming some of these problems in Fiji where the operation of the Native Land Trust Board has defused many of the uncertainties experienced elsewhere in Melanesia and enabled the creation of a market in customary land (Chapter 6). Local authorities in Fiji, though short of funds and expertise like everywhere else, have retained democratically elected town councils and generally possess sufficient skills to manage urban development. This is partly due to a more robust and diversified urban economy in Fiji, more adequate human resources and a longer urban history. Elsewhere in the region land problems are far more apparent, ranging from outright opposition to customary land registration in PNG towns, to a long backlog in confirming land titles in Micronesian countries. Although some common approaches are possible across the Pacific, priorities vary according to local circumstances and needs.

Third, all urban areas in the independent Pacific, apart from some mining towns (which often function as 'company towns'), are in a characteristically weak financial condition because of the lack of a significant private sector, the shortage of wage employment and the consequent absence of sufficient taxable income to support effective urban administration. This situation is compounded by problems in tax collection and a low level of cost-recovery that has made urban infrastructure provision an overwhelmingly aid-dependent exercise. There are no ready solutions to this dilemma; even those bodies that have looked at it closely, such as the World Bank (1995a), have merely suggested that more will have to be done with fewer resources in the future. Thus far this has barely discouraged the more grandiose urban development schemes, such as building urban freeways and parliament houses.

Fourth, some characteristics of governance in the Pacific have made the establishment of robust urban administrations peculiarly difficult. Effective urban (or regional) government in small island countries is constrained where national (and provincial) governments conflict and compete with urban councils. This is seen in the competition for scarce resources among the tiers of government, with local government losing out to those higher in the pecking order; in the shortage of personnel with political and technical skills; and in the domineering presence of national government located in the capitals. A further difficulty lies in systems of inherited colonial urban government that depend on voluntary and largely unpaid services of elected citizens to manage the government process. Councillors are expected to act in the 'public interest' and earn incomes elsewhere. Western urban planning systems generally assume that some popular participation in decision-making takes place and that the resulting plans reflect majority wishes to a considerable extent. These conditions are absent in the Pacific because of authoritarian formal political structures, limited awareness and the lack of consultative procedures. Greater public participation might follow the fostering and inclusion of non-governmental organisations, including religious and environmental NGOs, women's groups and the churches, in the determination of urban development.

Fifth, administrative difficulties face urban councils in conducting duties laid out for them by law and in government regulations. In addition to tax collection, there is difficulty in gaining cost recovery from the use of public utilities and services of a kind that were formerly supplied without charge or at a highly subsidised rate. The lack of investment in water supply systems in many towns is one consequence. Weaknesses in urban administration in many parts of the world, including the Pacific, have led to widespread introduction of corporatised or privatised services that are largely removed from public management. The most pronounced problem is that of implementing measures contained in urban plans that are only advisory in nature and not backed up by legislative sanction. This take-it-or-leave-it approach has clear social origins but prevents better management of urban development. Failure to enforce procedures is rarely due to a lack of powers but rather to an inability or refusal to use what measures are already available. Pursuing townspeople through the courts for contravention of land use planning regulations is generally implausible under circumstances where customary lands exist alongside controlled municipal areas (and courts and police services are fully stretched). Policing civil misdemeanours is not part of the 'Pacific way'.

Sixth, particular challenges to urban management and development are posed by the marine environments of most towns where avoidance of damage to fragile ecosystems is extremely difficult to ensure and exceptionally expensive to achieve. In spite of considerable progress in heightening environmental awareness in recent years, developing effective sanitation systems to ensure minimal pollution and safe waste disposal and the introduction of environmental advisory plans of various kinds are affected by constraints similar to those confronting land use planning systems as a whole. Environmental planning is only useful if integrated with the overall process of planning and development control and cannot exist as a separate branch of administration. There is little evidence of this integration in the Pacific or a likelihood that the state of the urban environment in particular will significantly improve until it occurs and planning offices are strengthened.

Successful development planning in the Pacific (and development control) requires a more sophisticated understanding of what good governance in the region might entail than is normally found in technical appraisals of urban affairs. Good governance involves legitimacy, and accountability, the ability to govern effectively and deliver public services and a form of coordination whereby 'political authorities work through, and with, a plurality of other actors [such as NGOs] and cannot be sharply distinguished from them' (Larmour 1995: 17). It must be linked to knowledge and expertise. The conditions for good governance are hard to find in some Pacific island administrations (as they are in the governments of many countries), hence establishing a viable planning system has been difficult.

One observer has suggested that in Micronesia at least, history and cultural environment have discouraged planning: 'why does planning not take place? Micronesians have neither the training nor cultural inclination for planning. The backstopping for planning has always been provided for them' (Ballendorf 1991: 3). It is generally accepted that modern urban planning originated in the need to deal effectively with deteriorating public health and sanitation during the industrial revolution in the cities of nineteenth-century Western Europe. Pacific towns, small scale and recent in origin, have emerged without high-impact industrialisation taking place. Only the contemporary realisation that basic transport, water supply and waste disposal systems cannot cope with the pace of urban growth in much of the Pacific is providing the preconditions for change. There is little historical context for towns, let alone urban planning, and the absence of long-term planning in the region (as economic development plans have been abandoned) parallels the broad lack of sustainable development in the Pacific despite assumptions about the long term environmental stability of customary wisdom (Connell 2001). Effective governance is largely promoted by external agencies like the World Bank and bilateral aid donors who are not involved directly in political issues, and hence meet opposition in post-colonial times; thus the PNG government construed the 1996 World Bank Structural Adjustment Programme as being little short of an attack on its sovereignty.

Pacific governments have experienced great difficulty in managing urban development and overcoming problems of unplanned growth. Maintaining an adequate quality of life in towns, which were generally established with all the technical and financial support of colonial administrations, has proved almost insurmountable in some cases, making attempts to anticipate and plan for future change something of a luxury. The movement towards more sustainable forms of urban living and development demand the existence of management systems robust enough to implement decisions about land use for the benefit of the whole community and act as a 'bridge' to overcome the many obstacles in the transition from traditional ways to modernity, yet few contemporary institutions are capable of this, even in the most modern urban core of developing island nations.

6 Land and urban life

> Land is our life. Land is our physical life – food and sustenance. Land is our social life; it is marriage; it is status; it is security; it is politics; in fact it is our only world.
>
> (Dove *et al.* 1974: 182)

Land tenure has been studied in detail in the Pacific, matching its importance for island life and economies, and thus especially in rural areas (Crocombe 1987; Ward and Kingdon 1995; Larmour 1997). The particular impacts of land matters on urban development have received much less attention, yet in both Suva and Honiara land was one factor in the constitutional upheavals of 2000 (R.G. Ward 2000). In part this reflects the recency and limited extent of urbanisation in the region but is also due to the failure of governments to develop successful policies for managing urban land. Customary land holding is the most widespread form of tenure in Pacific island countries, with much smaller areas being held under individual freehold or government ownership. The importance of communal sharing in land as long established in Pacific tradition is usually given as a primary reason why policy changes to facilitate individual forms of tenure have proved so difficult for post-colonial governments. Ward and Kingdon (1995: 14) have suggested, however, drawing on an earlier discussion by Hobsbawm and Ranger (1983), that: 'Invented traditions are common in the Pacific, and they have often been used as a key element in the establishment of government or other authority.' Traditions may not always be very old with respect to land, where critical changes were brought about by colonial authority or, in the case of Tonga, by an indigenous government in 1875 (Ward and Kingdon 1995). In some cases, as in Fiji, colonial governments acted to protect indigenous tenure by legal codification. Each particular form of tenure, and varied interpretations of them, has posed problems for changes in land use and ownership, and especially for the conversion of rural land to urban uses.

At least four crucial challenges face contemporary Pacific urban administrations, all relating to land. The first is the continuing strength of traditional attachments to land as a communal resource and the widely held conviction that traditional tenure must not be interfered with by governments or commercial interests. As Meleisea (1992: 64) has noted for Samoa, many elements of tradition (*fa'a Samoa*) continue to dominate social behaviour, even in the capital:

> Most people in Samoa continue to live in nu'u [locally based polities] and to practice a subsistence based economy. *Fa'a Samoa* provides the basic structure on which the subsistence economy operates. These rules guarantee that everyone, at least theoretically, regardless of sex, rank, social origin, descent or religion is assured of a piece of land to cultivate and live on, shelter and protection and care in their old age.

Broadly this is true throughout Oceania, though the strength of attachment to customary land tenure varies considerably from the near-universal support found in the 'village' communities encapsulated in towns, to a rather different situation in some of the more modern suburbs of Suva and Port Moresby, where residents occupy and own land to which they have no historic attachment (though they may be strongly attached to land elsewhere).

A second issue of growing concern is the basic need of relatively new commercial, mining and manufacturing organisations for certainty over land titles. Procedures for acquiring land ownership or registering commercial interests in land can be cumbersome, even in those cities where secure title can eventually be established. The situation in the FSM is familiar throughout and beyond the Micronesian region:

> An area that needs attention is the regulatory environment, especially the land lease market. The FSM constitution prohibits non-citizens from acquiring land, most of which is communally owned. In some states there are 25 year limits on leases. Most foreign investment projects require approval by both the national and state governments . . . these and other regulatory matters . . . amount to costly and time consuming barriers which in some cases may not be overcome by returns to investors.
>
> (Bank of Hawaii 1995: 16)

A general problem thus exists in 'adapting tenures which derive from combined customary and colonial precedents, to serve the needs of non-customary, post-colonial societies', whilst 'traditional precedents are not relevant to modern urban living' (Crocombe 1987: 390, 386), as demonstrated in PNG by the lack of a word in any local language to describe the sale of land. Pressures to reform traditional land practices are greatest in the urban centres. The impetus for change comes not only from local and foreign entrepreneurs, keen to use land as collateral for development loans, but also from major international agencies like the World Bank which may require the reform of land tenure practices as a condition for the approval of IMF Structural Adjustment Programmes, as happened in PNG in 1996. However, not only do such imposed reforms result in misgivings, but a modern land titles system is a sophisticated institution demanding high levels of administrative competence:

> Managerial skills of that order do not abound in the region, and such as there are, they are not immune from the manipulations of the wealthy and the powerful. Even if a Pacific nation did decide to change the customary system from one of kinship-based collectivities to one favouring entrepreneurship, landlords and tenants, rich peasants and landless laborers – to do so without adequate bureaucratic skills and strong magistrates is only to invite corruption on a massive scale, and no guarantee whatsoever of increased productivity.
>
> (Ward 1994: 47)

No Pacific nation has yet come close to making such a change.

A third challenge, particularly in the largest cities, is that of satisfying the land and housing needs of urban migrants. The structural or 'artificial' shortage of suitably located land for development in Port Vila and Tarawa, for example, has led to the spread of uncontrolled settlement at the urban boundary and into sensitive watershed areas, threatening the quality of urban water supplies. Yet extensive areas capable of being developed for housing, or other functions, exist within the urban boundary. The position of settlers is particularly difficult. Recent arrivals are vulnerable to exploitation by traditional landlords who control communal land holdings within and surrounding the urban area, where they are willing to lease land, or are forced to squat illegally on public land without the provision of basic amenities. Opposition to settlers, and urban residents generally, from the 'true owners' of urban land intensified in the 1990s, as both populations have grown, employment opportunities have dwindled, and crime and environmental degradation have increased. Chiefs and landowners in and around Vila have threatened those who fish, hunt and trespass in urban fringe areas, and their food gardens have been bulldozed, Port Moresby residents have constantly demanded the removal of settlers (Chapter 3), settlers have been attacked by dissident army personnel in Suva, and in Honiara a civil war emerged, in large part as a result of the concerns of local Guadalcanal people over the incursions of Malaitans on to their land, as migrants settled beyond the town boundaries, and gained access to their jobs. Land and urbanisation are intricately entwined.

Fourth, especially in Melanesia, all changes in land use towards the 'modern' have been complicated by sometimes enormous demands for compensation by traditional landowners or those claiming to be so. As land has become more valuable the number of claimants has invariably risen. Such claims are frequently made by customary landowners in Melanesia seeking monetary recompense for parts of their holdings given up or sought for development purposes, such as roads and other public and commercial uses, and there has been a rapid escalation in monetary demands in recent years (see, for example, Toft 1997). In PNG this has already resulted in closed hospitals, schools and airstrips, disrupted road construction and considerable violence, particularly in mining contexts. At the start of 2001, for example, two Goroka landowner groups threatened to cut off the town's water supply and sewage services if they were not paid K900,000 (US$300,000), since the payment of K1,500 in 1965 was considered inadequate (*PC*, 22 January 2001). A culture biased towards compensation has slowed the pace of social and economic development. As a Marshallese government official has put it: 'Once people said they belonged to the land; now they say the land belongs to them' (quoted in Hezel 1999: 8). Land issues confront every phase of contemporary physical planning in the Pacific, increasingly suggest the emergence of new forms of materialism (Connell 1997a), and impose costs and delays on many components of urban development.

In this chapter attempts by island governments to grapple with these interrelated issues are illustrated in the recent experience of urban land management practices. Effective urban management and planning demand an ability on the part of the urban authorities to control and regulate urban land use and set public priorities, but this is exceptional, not so much because of any failure on the part of Pacific governments to acknowledge it but due to a lack of agreement as to the most appropriate means of achieving it. Indeed there has long been a consensus that:

> Public authorities must ensure that [urban land] is made available at reasonable cost to all who need it and must avoid land becoming concentrated in the hands of a few persons who, by the accident that their ancestors held rural land which became urban, or because financial advantage has enabled them to acquire a monopoly position, derive large sums from less fortunate fellow men. This necessitates comprehensive planning and either extensive public ownership of land which is leased to users, or extensive public control over private land. The socially generated surplus in rising land values should be used for the common good.
>
> (Crocombe 1987: 387)

That situation has never been met with in practice.

Land in Melanesia

Underlying many problems attached to urban planning and service provision in Melanesia are land tenure issues involving customary areas within and around the towns, and the use of particular parcels of land for public purposes. There is an acute shortage of land in many urban areas; this became so severe more than a decade ago in Rabaul (PNG), for example, that it prompted a search for other sites for the town, a search that was unresolved by 1995 when volcanic eruptions destroyed the town and resettlement was inevitable. In Port Moresby extensive areas of traditional land within the city boundary are effectively quarantined from urban development, making service provision in fragmented suburban areas more expensive and more complicated. Land acquisition problems have prevented the National Housing Commission from acquiring building plots for housing in PNG towns, and the Waterboard from finding land for pumping stations, treatment plants and sewage disposal areas. There are widespread problems in finding suitable locations for garbage dumps, resulting in informal, illegal and environmentally damaging waste disposal, without support for compulsory land acquisition.

Urban customary land

All Melanesian countries place great store on the continuing role of customary land in support of national values, hence freehold land tenure is rare in Melanesia. Vanuatu went further than any other Pacific island country in limiting free market ownership of land by abolishing all freehold land ownership at independence in 1980 and returning it to traditional owners (Rodman 1995). Urban land issues are especially complicated by the structure of traditional ownership. In Port Moresby, for example, the Motu-Koitabu, the traditional residents of the district, are the sole owners of customary lands which constitute at least one-third of the city's total area. Moreover, some Motu-Koitabu villages, like Hanuabada, are actually incorporated into the National Capital District (NCD) area, presenting a particularly difficult task for homogeneous and orderly urban management. Land problems in Port Moresby are more complex and difficult than elsewhere in PNG because of its long colonial history, its location adjoining and around existing traditional villages, and the pace and extent of its recent growth. The land on which the city is built comprises government freehold, government leased land, private holdings and customary land; unused government land as well as customary land areas are scattered within the city boundaries, hence development proposals involve dealing with customary owners directly (Lea 1983).

Customary landowners in Port Moresby, as in other PNG towns, fear the gradual loss of ownership of their land, the historic basis of their security, and thus seek to protect it against further alienation. Landowners feel they have little control over migrants who have settled on their land because there are no effective agreements with them. Sometimes only one person in a particular land-owning group has gained by giving permission for migrants to settle, enabling the collection of rent from them, while other owners are excluded from any economic benefits. Tension and uneasiness often develop between migrants and indigenous landowners, especially as settlements grow and their permanence becomes more evident. Confrontations have taken place but neither in Port Moresby nor anywhere else have squatters forcibly occupied land against the wishes of the traditional owners. However, as time has passed and settlements expanded, traditional owners have sought to repossess their land or secure new rental agreements from settlers. Yet there is no easy way to remove migrants, should landowners wish to do so, and migrant groups tend to grow in number, size and authority, after a period of residence, by dint of their growing power and status in urban society.

Land issues are further complicated by inevitable uncertainties over the specific nature of group ownership of particular tracts of land, especially when such urban areas have been alienated from their traditional owners for many years, and by indecision and uncertainty amongst traditional owners concerning the future of the land. At Blacksands settlement (Vila) some landowners lived in the settlement whilst others were more distant; more distant landowners wished to legally subdivide the land to lease to the province or possible investors, others sought to increase the number of settlers paying rent, some preferred the removal of settlers and were opposed to improvements to the infrastructure (fearing that residence might become permanent), whilst still others had no particular plans for the land (Mecartney 2001: 92). All were conscious that their need for the land was likely to grow in the future, but there was no organisation, formal or informal, through which they could articulate sentiments, fears and plans.

Over time settlements acquire the characteristics of suburbs, with access to services (often on the strength of votes that can be bribed and bought), including roads, electricity and water, churches and stores. After less than a couple of decades Blacksands had six churches, two bakeries, ten stores and ten kava bars, two kindergartens and a dispensary to support a population of around 400 households. Yet there were just two water tanks (and many polluted wells), no piped water supply or telephone and no primary school. Sewerage systems are inadequate and ineffective and contaminate the groundwater. Solid waste is burnt and the fires have become a health hazard. Some residents, through dint of early arrival, had become minor landlords, constructing and renting rooms in their own 'yards' within the settlement (and paying their own rents to the local landowners from nearby Ifira island). Despite the semblance of permanency attached to the acquisition of services and the gradual improvement of housing, residents had no security of land tenure. Acquisition of secure title to land was the greatest need for most residents, resulting in unwillingness to invest in housing or infrastructure, so reducing the quality of life (Mecartney 2001). Similarly, that same insecurity discouraged the extension of urban planning to Blacksands (though there was little enough of that elsewhere). Insecurities of tenure were problems for settlers, landowners and urban managers.

Commercial leasing of customary land for business ventures or any kind of economic or industrial development is difficult. As a group landowners are often unable to reach agreement because of their conflicting interests, as at Blacksands. Customary land cannot be used as security for borrowing and banks cannot take over customary land if a borrower defaults. As a result, the Motu-Koitabu, for example, could not take advantage of their most valuable asset and have not benefited from the rise in values that location in Port Moresby has brought (Morauta *et al.* 1986). This has led to their elected representatives in the national parliament occasionally recommending that they demand more. In 1992, Jack Genia, then the Pangu Pati Deputy Leader, stressed that the main beneficiaries from urban development were not the traditional landowners, but companies and even the churches. Genia noted that:

> Some churches have acquired so much material wealth through ownership of some of Port Moresby's highly valued portions of land and properties.

> . . . When churches should be helping to overcome social ills of the displaced people, they are investing millions of kina in real estate properties which share boundaries with displaced landowners living in shanty towns and settlements.
>
> (*PC*, 25 September 1992)

He urged the Motu-Koitabu to fight hard for their rights, reclaim unused lands now occupied by others, and claim compensation for their displacement within the city. However, few administrative mechanisms can help the Motu-Koitabu landowners exert their rights and claims, causing them enormous frustration. At the end of 1992 they sought removal of all those living illegally on land in the capital city; a meeting of customary landowners indicated that many had reached 'boiling point' at seeing their land occupied by numerous illegal and unemployed squatters (*TPNG*, 30 December 1992). Despite there being a Motu Koita Council, the absence of formal mechanisms to deal with this perceived problem and the absence of effective democratic processes in the capital prevented responses other than incomprehension, despair and occasional violence. Little had changed by 2001 when the Chairman of the Council claimed:

> We, the Motu Koitas, have lost our gardening land, our hunting and fishing grounds; we have lost our employment and business opportunities, our school spaces and our health services; we have lost the freedom to move around in town without fear; we have also lost our culture and tradition. And worst of all, if we are not careful, we will lose our destiny, our heritage and our identity. . . . We have become spectators of development on our land.
>
> (*PC*, 8 January 2001)

In some situations, however, arrangements have been worked out among the urban residents themselves: 'In the peri-urban areas particularly, the original rightholders will say they have been elbowed out by the aggressive immigrants; the latter will say they have been mulct and then threatened with expulsion' (Ward 1994: 48).

Although in both Port Moresby and Lae some formal associations of established and migrant groups have worked out leases with the assistance of state officials and urban planners (ibid.), this has usually proved impossible. In Lae the residents of local villages engulfed by the city formed themselves into the Ahi Association, first, to mount a land claim – that had persisted for half a century – to land alienated by urban development; second, to press their involvement in discussions of development that affected their land (and from which they had been excluded), and to oppose any plan contrary to their interests; and third, to secure more formal rights of land tenure against the incursions of squatters, who were increasingly perceived as encroaching on scarce land. The Association was without formal political representation at any level, and unable to secure through the courts what it regarded as a satisfactory outcome, hence it was ignored by the Huon Council (which covered Lae and the lower Markham valley). It therefore mounted a determined effort to remove unwelcome (unemployed) squatters by serving notices to quit and destroying their gardens, but without political power these actions had little effect. Individual villages within Lae also formulated their own development plans, advancing 'the ludicrous situation where the constituent village bodies are now moving to a situation where they will each have their own separate development plan' (Adams 1982: 229). Ironically two decades later Ahi claims to Lae land were disputed by the nearby Wain people (*PC*, 9 October 2000), one indication of the rising value of urban land. All this occurred in a context where attitudes to urban development varied considerably between the conservationist views of the incorporated villages, the demands for services and improved housing from settlers and the attempts by councils to impose urban development plans without consulting either group. Though the Ahi Association – like the Motu Koita Council – strengthened local corporate identity, it could not contribute to improved urban management or the resolution of land issues because of the different interests and scales that were involved. Without resolution at the urban level, both organisations accentuated tensions rather than resolved problems.

Such situations make urban planning difficult, delayed and unusually expensive. More generally the combined effects of population increase, declining rural incomes, land degradation and diminished opportunities for both rural and urban villagers have resulted in frequent demands on government for compensation. Paul Wanjik, the Minister of Works, has commented:

> It is regrettable that land problems or to be more specific compensation demands are fast becoming a growth industry of their own. People are now turning to land compensation demands as an easy way of making a fast buck. This behaviour is fast reaching epidemic proportions. I fear that if nothing positive is done to control and regulate this, we will experience some very serious problems in implementing capital works projects.
>
> (*TPNG*, 27 December 1990)

Retrospective claims add extra uncertainties, but the legitimacy of landowners' rights and the significance of land to livelihood weaken challenges to these claims. The combination of disputes over land ownership, increased claims for compensation by those who often perceive this to be their most valuable source of income, and mismanagement have all slowed the process of urban development in PNG and rendered it more complex and expensive than in most other countries. One outcome of this is that in the larger urban centres, and especially Port Moresby, a higher proportion of urban residents than anywhere else in Oceania are in squatter settlements.

Urban villagers, such as the Motu-Koitabu in Port Moresby, may also be disadvantaged in ways other than land alienation. Although often perceived as being well-off by having free land close to the city, gardens located nearby and good jobs in town, the real situation is somewhat different. Those who live in urban villages include some of the poorest in town; they tend to have larger households, lower incomes per resident and more households without wage-earners than the urban average. Their position is very similar to those living in migrant settlements because their poverty is directly related to the fact that their land is in town. They are, after all, permanent urban residents and have nowhere else to go when they are old, poor, sick or unemployed (Morauta *et al.* 1986); unlike most urban residents they cannot ultimately take advantage of the rural 'subsistence safety net' that is still more or less in place throughout the Pacific. Every sick or elderly Motu-Koitabu lives in the city, as do the unemployed youth and children with no place in school; moreover the majority are unemployed (*PC*, 29 September 1992). Their land is little help to them, since they have lost local fishing grounds and nearly all convenient land suitable for agriculture. The ten villages within the NCD are also poorly provided with urban services and none of them has an easily accessible clinic. Although there are community schools in Motu-Koitabu areas, there are too few places to accommodate all school-age children. Most urban villages are very densely populated, leading to health hazards and untidiness, and recreation facilities are usually lacking. Ironically these are very similar to the conditions experienced in the migrant settlements and the needs of both groups have not been met, and cannot easily be met, to the same extent as are those of others within the city.

Responses to the land dilemma in Melanesia

Land mobilisation strategies have been pursued in PNG which would provide both urban and rural land suitable for development through the registration of customary lands. The introduction of a World Bank-funded Land Mobilisation Programme in 1988 was principally aimed at creating a clear policy and legislative environment to promote investment in land, enabling customary landowners to participate in the development of their own land, and transferring land administration responsibilities from the national government to provincial governments. When the programme was initiated there were considerable difficulties to be overcome, including 'ineffective management of the Land Mobilisation Programme, inability to coordinate and absorb technical assistance, the quality of staff that can be retained or recruited into key positions' (Papua New Guinea. Department of Lands and Physical Planning 1988: v). Since then these difficulties have become more apparent as it has proved almost impossible to mobilise more than a small proportion of the land required for urban development.

Some progress has however been made in PNG with securing finance for land development by introducing the state as an initial lessee of customary land. Lease–lease-back schemes for customary land mobilisation, using the state as lessee, have enabled both landowners and developers to obtain finance for development. In towns like Lae, Wewak and Alotau, landowners have been allowed to sell or lease land for medium-term development. The intention was to bring customary land adjoining urban areas into use without it becoming alienated government land. Another scheme known as 'direct dealing' has permitted owners to lease land to PNG nationals in accordance with custom and without formal registration (Asian Development Bank (ADB) 1992). Despite such varied schemes a host of institutional problems have overwhelmed the agencies and individuals responsible for administering land reform. Complex legal and administrative processes for allocating leases, institutional weaknesses, including shortage of management and professional skills, and the absence of a clearly defined policy created backlogs and delays in lease allocation to investors. Coordination between government agencies was weak and disputes exceeded the capacity of the mediation mechanisms to resolve them (Papua New Guinea. Department of Lands and Physical Planning 1988: 7).

Appropriate compensation and leasing regimes for land that will be alienated for particular periods of time and different purposes remain essential, especially since there are differences between the absolute loss of land to urban development or mining and its temporary utilisation by forestry and plantations. In PNG most compensation has been *ad hoc*, worked imperfectly and constantly been challenged by the supposed beneficiaries (Toft 1997). Both rural and urban people have, in some cases, bargained away their land resources, for short-term gains, in a manner that could prejudice their future survival (Connell 1997a). In a sense, that occurred in early decades of the twentieth century within urban areas. Although landowners have sold such physical resources as timber without compensation and for limited material gains, there is enormous resistance to the alienation of any more land.

After many official inquiries and reports, the PNG government with encouragement (and some coercion) from the World Bank decided to proceed with the preparation of national legislation to make registration of customary land possible. A proposed Land Registration Act was still under investigation when outbreaks of community violence in Port Moresby in 1995 were directed against registration. Several political parties and student groups opposed the inclusion of land mobilisation as one of the twenty-seven conditions required by the World Bank in their Structural Adjustment Programme (SAP), raising their concern that a landless class would be created but more particularly that land would no longer be controlled by the traditional owners. The SAP was only agreed late in 1996 after the land registration condition was removed. Issues of land tenure had come close to bringing down the government.

Apart from Fiji (see below), there is a marked reluctance on the part of governments in Melanesia to intervene in customary land practices, leading to codification:

> generally the instinct of the political majorities is to pull back from major changes, to leave the formal systems much as they are, and to find what flexibility they can within the customary system, or the small proportions

> of land which were acquired by imperial regimes and are now within the control of independent governments.
>
> (Ward 1994: 47)

In the Solomon Islands, for example, there is relatively little customary land in the urban areas. Part of Honiara was already alienated for a copra plantation before becoming urbanised; this and other areas were compulsorily purchased by the colonial government in 1947. There have been continuing disputes however with peripheral landowners about areas for water supply, garbage disposal and a proposed dam behind the town, and ultimately over the necessary expansion of the urban area. Honiara benefited from the past alienation of a substantial area of land in a sparsely populated area, but increasingly the situation has become much like that in PNG. Indeed the expansion of settlements beyond the officially alienated area, intensifying disputes with local landowners, was a critical influence on the rise of ethnic violence in 2000.

At independence all alienated land in Vanuatu, including government holdings, was returned to its original owners. The Constitution allowed special provision to be made for urban land and a policy was adopted that would allow the government to 'declare' that some customary land would become 'public'. It would then be managed, but not owned, by the government or by statutory bodies set up for the purpose. The customary owners would also be involved in the management and share part of any revenue raised. The tenure system was thus rather like a compulsory lease, with the rents paid to the customary owners being related to the revenue collected. Port Vila, with its boundaries slightly amended, was declared 'public land' on Independence Day, and a Vila Urban Land Corporation (VULCAN) set up to manage it. The corporation was expected to create substitute leases to replace the freehold titles removed at independence: 'In urban areas, land titles were converted on independence to automatic leases which, however, rested on the notion of customary ownership and on the assumption that rents from leases would pass, through authorised bodies, to custom owners' (Rodman 1995: 83). It was a bold step that could have led to a collapse of foreign investor confidence and capital flight. In practice, however, the considerable political influence of the customary landowners of Port Vila (the villages of Erakor, Mele, Pango) altered these arrangements. The policy was amended so that all rents raised by government from public land, not just a share of them, would go to the owners after the administrative costs of VULCAN had been covered. However, it could not be decided how the money should be divided up between the three villages, with the result that five years after independence large sums had been accumulating unspent in VULCAN's bank accounts with the landowners undecided how it should be shared (Halliburton 1992).

The urban land situation remains somewhat confused in Vanuatu after accusations of corruption led to the disestablishment of the Urban Land Corporations both in Port Vila and Luganville. After five years of concern the Minister of Lands dissolved VULCAN in 1988 and took direct control of its operations. Urban villages took this 'as a direct attempt to terminate recognition of their customary rights to Vila town' and 3,000 demonstrated in protest (Rawlings 1999a: 76). In the early 1990s the Ministry of Lands forecast that no further land would be available for urban development within the urban boundaries of Port Vila for at least a decade, even though there were extensive undeveloped areas throughout the municipal area. The ensuing shortage reflected customary constraints over land supply rather than any physical deficiencies (Chapter 5). Both land developers and those seeking housing moved to the fringes of the city where accessible and available customary land existed. An agreed development procedure was established via a 'negotiator's certificate' granted by the Minister of Lands which sanctioned the terms of leases with the customary owners (Halliburton 1992). However, in common with many land disputes in Melanesia, this procedure demanded agreement as to who the customary owners actually were and the nature of existing interests in the land, but that proved too difficult. Acquisition of customary land in the public interest is legislatively possible in Vanuatu but examples of this are rare. Gaining title to urban land remains virtually impossible.

Land registration in Fiji

In Fiji special provisions were introduced early in colonial times – the late nineteenth century – to safeguard indigenous use and ownership of customary land; subsequent codification has ensured a complex system, quite different from that found in other parts of Melanesia, that provides access to indigenous lands for development purposes (Overton 1992; R.G. Ward 1995). Land was divided into privately owned freehold areas granted to foreign settlers, Crown (government) lands, and 'native' lands; these proportions are now approximately 10 per cent (freehold), 8 per cent (Crown land), and the balance of 82 per cent (Fijian native lands). Unlike elsewhere in Melanesia, Fiji has private freehold land that can be bought and sold by non-citizens. Suva, with a population not much smaller than that of Port Moresby, is built on both private freehold land and Crown land and incorporates several Fijian villages located on native land. Customary land in Fiji is managed quite differently, nearly all being registered with a *mataqali* (variously defined as family or clan) as the unit of ownership. Nevertheless, registration is still incomplete and it has required an enormous bureaucratic effort to maintain the records of the Native Land Commission. The fact that land registration remains incomplete in Fiji – a geographically small (though fragmented) area of minimal regional cultural variation – is indicative of the huge task that confronts registration in the other Melanesian states.

Since 1940 the rights of indigenous Fijians to sell and lease their lands have been compulsorily transferred to a Native Land Trust Board (NLTB), a statutory body managed and staffed almost completely by indigenous Fijians, which acts as the agent of some 6,600 traditional landowners. Native land is divided into two categories: 'reserved' land, judged to be needed by the owners for their own subsistence, and 'unreserved' land. If the Board wants to lease reserved land, it must get the written consent of the owners, but it can lease out unreserved land for up to 30 years without their consent (though in practice it consults them). The Board has proved an expensive institution to run, despite automatically deducting 25 per cent of income received from leases to cover its administrative costs. This revenue has not been enough to maintain its activities and it has had to seek additional government grants, reducing its autonomy. Payment of rents to landowners supports the Fijian chiefly system and chiefs automatically receive a percentage of the rents. After the Board and the chiefs take their automatic share of rents less than half is left for distribution among ordinary members of the *mataqali* (Nayacakalou 1971).

The NLTB was set up as a way of making land available for development while simultaneously protecting the interests of its Fijian owners. To do so it actually restricted native Fijian rights to sell or lease their own land (UNDP and UN Centre for Human Settlements 1991). At different times the NLTB has been perceived to be a solution to urban land problems in other parts of the region, especially for facilitating the supply of customary land in towns. Thus the Mount Hagen Town Authority in PNG, in its 1990 Annual Report, recommended the establishment of a Native Lands Trust to administer land on behalf of the landowners; whilst the Samoan Cabinet prepared a new Land Registration bill in 1992, 'which would facilitate the development of customary land by channelling leases through the government, in much the same way the Native Land Trust Board operates in Fiji' (O'Meara 1995: 153). Such developments in neighbouring states suggest that the Fijian system might be beneficial for urban administrations throughout the Pacific, yet its success derives from the legal recognition of customary lands a century ago and its codification in authoritarian colonial times. Other countries in the Pacific have tried to address the customary land situation after independence when different circumstances and changed legal frameworks have made the Fijian model, or something approximating it, far more difficult to implement.

Land in Polynesia

Economic individualism is more evident in the largest Polynesian countries of Tonga and Samoa, alongside claims of individual ownership over customary land (James 1995; O'Meara 1995). None the less, urban land for public purposes is scarce, especially in Samoa, because governments are unwilling to exercise their existing legal rights to acquire additional customary or individually held land for municipal uses. At the same time, working through traditional avenues is not an effective option because chiefly systems which were formerly so important in the allocation of lands have weakened, resulting in severe pressure on the small amounts of urban land that do exist and little opportunity to extend or redevelop existing residential areas. This has necessitated reclamation of marine and lagoon areas for public use rather than attempts to expropriate land from customary owners. In Tonga, rather than the government initiatives of Samoa (and also Tuvalu), private individuals, living on the physical (and social) margins of urban society, have reclaimed swamp and lagoon land for residential use. In Polynesia there is not so much failure to accept the need to acquire land for public purposes as lack of agreement over the most appropriate means of achieving it.

Land issues in Polynesia are no less complex than in the rest of the Pacific, but each state has taken a distinctly individual approach to land management, under circumstances where no ready solutions to physical shortages and cultural constraints exist. The land tenure system of Tonga is quite unique within Oceania since, unlike other systems in the regions, it does not simply provide for individual ownership but largely guarantees it. Tonga was never colonised and the Tongan constitution of 1875 did not permit the alienation of land to outsiders. Land was divided into royal, government and noble estates and all males were granted individual rights to land. All land is ultimately the property of the Crown but large estates are owned and operated by a small group of fewer than forty nobles and *matapule* (spokesmen for and executive officers of the chief) (Maude and Sevele 1987: 121). Every male Tongan on attaining the age of 16 years is entitled to an agricultural plot known as a tax allotment (*'api tukuhau*) of about 3.3 hectares (on which rent is payable) and a town allotment (*'api kolo*) of 0.16 hectares. The allotments are granted from the estate of the noble where the applicant lives, subject to the agreement of the noble concerned, and remain in the possession of male heirs as long as rent and taxes are paid. Land grants may also be made from land under government ownership. In theory this system provides for considerable equality of access to land.

Land and development in Nuku'alofa

The land system in Tonga was developed at a time when there was sufficient land for all adult males. It does not now 'conform to the generally accepted principles of town and country planning' (Tonga Central Planning Department 1991: 76), since little centralised and public control over the management of urban land use exists. Moreover, nineteenth-century land legislation was designed to prevent the appearance of a land market and was only changed in the mid-1970s, to allow leasing of customary allotments. The small size of holdings, however, and short lease periods (since few owners wish to enter into long-term formal leases) mean that informal arrangements are more common (James 1993). This in turn is not conducive to the emergence of regulated private sector development or the investment of development capital by either landholders who are not using the allotment or potential users who have no security of tenure.

Residential land in Nuku'alofa is in short supply, with newcomers forced to settle or squat in swampy areas around Popua and Sopu to the east and west of the town. Leases in the better parts of the town are expensive: T$150,000 (US$110,000) was asked in 1991 for a 10-year lease of 0.12 hectares in a central Nuku'alofa location, and T$30,000 (US$22,000) for a 50-year lease on a suburban site; both represent amounts difficult for local residents to afford (James 1993: 225). Tongans resident overseas are increasingly taking advantage of high rentals available in Nuku'alofa to build new commercial dwellings to let on their own *'api kolo*. In the absence of enforceable land use planning, urban land uses are haphazard and ill defined, compounding the proper servicing of business premises and residential areas and leading to the proliferation of conflicting uses in neighbouring plots. The government has placed great store on the possibility of further reclamation of the Fanga'uta lagoon area to create new supplies of urban land, whilst at the urban margins rural allotments are increasingly being subdivided to provide town allotments.

Tonga has no urban land strategy. The *Sixth Development Plan* (1991–5), the last in the series, put the situation succinctly: 'Although more than a third of the population is urban, Government's policy towards urbanisation and human settlements in urban and densely populated areas has not yet been formulated' (Tonga Central Planning Department 1991: 257). Although the plan intimated that both a land use plan and environmental legislation for Nuku'alofa were under preparation, they never eventuated. As the *Development Plan* also stated: 'The absence of land-use planning and building codes hinders national housing development. In particular, there are no adequate zoning procedures or specific building requirements for construction within flood-prone areas' (ibid.). The advent of urban land use planning in Tonga would signify a major change in the amount of power exercised by traditional authority. In a hierarchical society, without local government structures, urban development policies are not imminent. Even the introduction of new procedures for the control of land development will ultimately depend on government willingness to implement them. Tonga, with an admirable and equitable overall land tenure policy, has not managed to adapt it to accommodate the changing structure of settlement since 1875.

Development pressures in Apia and Funafuti

Samoan land is made up of some 80 per cent in customary tenure (primarily rural but also in the Apia urban area), 16 per cent as government lands, and 4 per cent privately held under freehold tenure, mainly in Apia and its vicinity (O'Meara 1995). Control over land use in Apia is the shared responsibility of customary chiefs (*matai*), central government departments and agencies and the churches. What was true two decades ago remains true:

> The Apia town area is the dominant administrative, commercial, and service centre of Western Samoa, but it is also a collection of locally autonomous villages organized by descent group memberships and hierarchies of chiefly titles . . . *'Aiga* links are activated for securing residence and land rights. Chiefs control access rights to scarce residential land.
>
> (Tiffany 1979: 166)

In the traditional system land is owned by the extended family (*'aiga*) whose members first cleared and planted it. Changes associated with population growth, urbanisation, more cash cropping and alienation of some land have been accompanied by altered perceptions about land tenure. Samoans, like many other Pacific islanders, increasingly seek exclusive control over their own land rather than sharing those rights with extended family members (Thomas 1984; O'Meara 1995). Land in any particular place may be owned under old, new or altered tenure arrangements, although the new form of individualised tenure is not the legal norm, causing problems in resolving difficulties before the Land and Titles Court. Under the Samoan constitution it is illegal to sell or mortgage customary land unless authorised by Act of Parliament.

In 1984 land in Apia suitable for urban development comprised a mixture of four main kinds of tenure: freehold (1,948 hectares); customary (744 hectares); WSTEC plantation lands (719 hectares); and government land (437 hectares) (Western Samoa Department of Public Works 1984). Thus about half the land suitable for urban development within existing urban boundaries was freehold, and only 10 per cent was controlled by the government. The situation has subsequently changed very little regarding tenure but new construction throughout the urban area has stimulated a change in attitudes towards customary holdings. The 20 per cent of land held under customary tenure is almost all in residential use in the urban villages and thus subject to traditional control. The great majority (about 70 per cent) of housing in the urban area is built on freehold land but potential housing plots are now in very short supply, hence there are acute problems in finding land for housing development (EDU International in association with GHK/MRM International 1992b).

The presence of culturally distinct groups of people, in Apia as elsewhere, demonstrates that urban land holdings can be complicated by factors associated with early colonial interventions, and that social relations evolving from the nineteenth century remain crucial to urban development. In the mid-nineteenth century a German trading firm recruited Niueans, Polynesian migrants from the small island of Niue, to work in Samoan copra and cotton plantations. They were given permission to occupy land near the centre of Apia, known as Aai-o-Niue (Niue settlement), and more than a century later their descendants have extended their occupation to five blocks in the Tuamasaga district (Matheson 1987) originally divided according to which villages of Niue the residents came from (Samasoni 1987). The land is held under a mix of tenures, with individual rights of ownership certain in only three blocks. Uncertainty about security of tenure has prompted many Samoan Niueans to migrate to New Zealand (as have the majority of Niueans from their home island). Continuing doubts about the status of Niuean families living in Apia, despite the presence of a community for more than a hundred years, emphasise the strength, persistence and exclusivity of Samoan traditional custom surrounding land holding. The Niuean community must continue to act together to seek permission from the local high chief for the proper surveying of plots and the negotiation of continued leasing arrangements (ibid.: 89–90). The significance of cultural differences, even in Polynesia where urban areas are more homogeneous than in Melanesia or Micronesia, hampers a consensus over the structure of urban development.

The trend towards individual tenure has brought with it considerable distortions in the freehold land market in and around Apia. The minimum residential plot size is about a quarter of an acre because of sanitation requirements. Undeveloped freehold land is in short supply and prices for centrally located plots have risen sharply in recent years. Subdivided former plantation plots in the Siusega area, which had an original purchase price of WS$6,000 (US$2,500) in 1988, were worth WS$15,000 (US$6,000) four years later, or three to five times the annual take-home pay of a mid-level wage earner (O'Meara 1995: 152). Similarly, in 1994, half-acre blocks at Afiamalu and Aleisa were on sale at between WS$16,000 and WS$20,000. Inflated and rapidly rising prices have been obstacles to housing construction and business development and have created new planning problems because owners have had to build wherever they held land (O'Meara 1995). Residential land availability was further reduced by the failure of owners to secure development finance and by speculative land holding. The demand for freehold and *de facto* 'ownership' of customary land by increasingly nuclear families, even set against the continuing strength of *fa'a Samoa*, led in 1992 to a government review of the existing law that would allow leases on customary land to be extended to 99 years and so provide adequate security for mortgage loans and more opportunity to develop customary land stocks.

Land tenure in Tuvalu differs little from elsewhere in Polynesia since almost all land is privately owned apart from a small amount of communal land controlled by the government. Crown or public land is limited to foreshores and roads, and most government buildings are on leased land. About a quarter of Funafuti's land area (59 hectares) has been alienated or leased by the government, and indigenous land rights were codified and registered under colonial land commissions in 1935 and 1948 (Connell and Lea 1995: 89). Customary tenure has been simplified, landholding groups are smaller – with the household increasingly the most common unit – and land ownership is consequently fragmented. In the urban area of Funafuti there is an increasing land shortage, and land has acquired a significant monetary value; yet precisely because of shortages there has been a growing reluctance to sell, lease or transfer land, leading to more disputes (Leupena and Lutelu 1987: 165). Moreover, despite codification and registration:

> The push to develop the infrastructure of a modern urban . . . government and economy has tended to bypass land management and planning issues. The land and land-based resources of Fongafale Islet are now under severe stress and deteriorating at an accelerating rate: they are not even mapped so they can be studied, considered, planned and managed for appropriate protection, conservation, enlargement, enhancement and use for future

> generations. Much recent development and currently proposed development is environmentally unsustainable, as well as being economically unsustainable in terms of recurrent costs and the foreseeable availability of capital.
>
> (Clarke 1993a: 19)

Fragmentation and the absence of formal documentation of ownership are obvious constraints to urban management and planning. Even in the smallest capital city in the region, conflicts over land are no less serious than elsewhere.

Despite some minimal erosion in the authority of Polynesian leaders – and rather different leadership structures in the three states – to a very considerable extent traditional attitudes to land prevail throughout the region. Chiefly status is recognised primarily in terms of custody and control over land, particularly in Tonga and Samoa; government challenges to that chiefly authority have been particularly weak, not least because governments are primarily composed of nobles and chiefs. As a consequence urban administrations have struggled unsuccessfully to gain effective control over land, and urban planning has been conspicuous by its absence. It is no coincidence that in each of these states there has either been reclamation of marine land or the serious consideration of it.

Land in Micronesia

Although the Micronesian countries of FSM, Kiribati, Marshall Islands and Palau possess a great diversity of geographic features and land tenure characteristics, they share some common regional attributes in land development. Each experienced a lengthy but distinctive colonial period and thus acquired foreign administrative structures and legal systems that have strongly affected land use and tenure. The differences are reflected in diverse land administration issues and legislative controls. The atoll states of Kiribati and the Marshalls are clearly different from the high islands of FSM and Palau. Land is at a greater premium in the atoll states and urbanisation is characterised by very high density development, extreme pressure on urban services and problems with waste disposal. Pressure to accommodate outer island and rural people in the towns is a general problem throughout the region, hence demands for more rapid land reform focus on the urban centres. Throughout Micronesia social and economic changes have altered the way people regard land and weakened traditional ties with it in varying degrees. In some places, for example, the lineage no longer acts as the main social and economic unit, changing family relationships to the land (Hezel 1994), though the basic principles of land tenure have been largely unchanged for centuries. There are marked differences in the nature of land issues in the Micronesian capital cities related to the individual history, culture and economic conditions of each country. This diversity is illustrated here in the recent development experience of the three regional capitals of Kolonia (Pohnpei State, FSM), South Tarawa (Kiribati) and Koror (Palau).

Land development in Kolonia

Authority over land is very much a state responsibility in the FSM and the federal constitution largely ignores it. However, some of the most intransigent problems in recent years, such as local opposition to the presence of migrants from other parts of the country in the Pohnpei State and national capital of Kolonia, have federal implications and are unlikely to be settled by unilateral action in one state. Indeed, even that possibility is rather remote as the Pohnpei State Department of Lands has been likened to a 'toothless dog which can bark its policies and regulations but cannot do much more than bark because the executive branch has not been able to enforce any of the policies' (*FSM News*, 1 January 1996). Given the considerable economic differences between the four states this has guaranteed the persistence of a particularly complex land development situation.

Pohnpei is the more economically developed state in the FSM, although not the largest in population, and has the best supply of land suitable for both urban development and agricultural uses. The proximity of Kolonia to the new national capital at Palikir, a creation of independence, provides the preconditions for the possible growth of a relatively large urban centre, but land issues make this implausible. The land situation in Pohnpei is quite different from that in the other states. As much as 60 per cent of the main island was declared 'public lands' during the German colonial period (1909–14) but efforts have been made since the establishment of the American administration in 1947 to reduce these holdings by transfer of title or leasing to private owners. The patrilineal inheritance system imposed by the German land programme of 1912–14 was modified in the inter-war years by the Japanese to a matrilineal system, before reverting to both male and female inheritance introduced by the Americans in the 1950s, resulting in a highly complex and confusing mixture of rules (EMPAT 1996).

Pohnpei has greater control and responsibilities over land than other states in the federation because of the extensive areas of public lands that were designated in colonial times. Considerable parcels of this land exist in and around Kolonia and have become the target for squatting by Pohnpeians from other parts of the main island, migrants from the outer islands, FSM residents from other states, and even some foreigners. According to the Constitution, only native Pohnpeians can own (have a certificate of title to) land, hence local concern (in neighbouring Nett municipality) over urbanisation:

> if anyone would care to look around they will find that people from all over FSM are settling and homesteading on public lands inside Nett Municipality, even Filipinos. No action has ever been taken to ensure that the Nett natives get their constitutionally guaranteed right to settle first before letting others squat . . . As a result of this mass squatting the native people of Nett are being displaced by an entire generation.
>
> (*FSM News*, 1 January 1996)

Whilst scarcely true, movement of squatters on to public land can hardly be avoided in circumstances where wage employment is so concentrated in the main town and authorities are reluctant to evict their own citizens. Squatters have moved faster than the institution of land registration to the extent that some squatters have been able to 'sell' land on which they have settled for as much as US$10,000 for 970 sq m. The registration and transfer of this public land would ameliorate the situation and prevent disputes over tenure becoming a political issue.

Numerous problems face the state government in coping with land issues, including difficulties in completing land registration begun during the American administrative period. Many boundary disputes arose because of faulty plot identification during the process of public land transfer which held up the transfer of other public lands (such as Japanese and Navy leaseholds) to private title. The extensive backlog of surveying and mapping delayed the supply of suitable land for commercial leasing. Land speculation is not yet a major problem but bottlenecks in land supply and the presence of a landed class of wealthy families who are also active in government and commerce point to the likelihood of future difficulties in achieving land reform.

Land development in South Tarawa

All land in Kiribati is privately owned, other than a small amount of government land comprising airstrips, causeways, water reserves and reclaimed areas, and land has increasingly become a commercial commodity in urban South Tarawa. In some areas absentee landownership is a problem and in others the fragmentation of holdings into tiny allotments following subdivision hampers further development. Though fragmentation has provided a choice of residential sites, for almost 30 years administrations attempted to encourage land consolidation but without much success. Attempts at major reform, including that of complete registration, have been tried but never completed; as elsewhere, the contemporary challenge is to reconcile the complex tenure system with a legal land management process that ensures orderly supply of land for development. Responsibility for this is however unclear, with various agencies being involved including the Ministry of Home Affairs and Rural Development (KMHARD), the Central Land Planning Board (CLPB), the Betio Local Land Planning Board (BLLPB), and the three local government councils within the urban area. Alongside familiar fragmentation of endeavour and responsibility the land planning boards have no budget or other resources for implementation purposes and are thus reliant on other agencies. The institutional problems of land management thus reflect the uncoordinated nature of government administration of urbanisation.

The urban area of South Tarawa is less than 16 sq km in area and the forecast demand for building land (from the late 1990s to the year 2010) is of the order of 2,000 plots. Increasing demand from the growing population and the prevalence of informal land transfer procedures have led to pressures for more secure and individual tenure arrangements and more squatting. The government has long been reluctant to move against squatters because of the importance of traditional rights and is also severely hampered in preparing land use plans, even for its own public land (usually leased from customary landowners), because of the veto rights of landowners on all forms of development.

Land concerns are closely linked with several other important areas of public policy including the control of population movements to South Tarawa from outer islands, housing and the provision of land for public purposes. As in Polynesia, reclamation has long been a strategy to provide additional government lands because of high leasing costs of private land and the unavailability of land at some strategic locations. In Kiribati estimates have been made that 6.6 sq km could be reclaimed, but at a cost of A$82.6 million (US$60 million) (UNCHS and UNDP 1994), a sum well in excess of the resources of its small economy, but that this has been seriously considered is an indication of substantial overcrowding and the difficulty of land reform. In any event, more recent predictions about the impact of the greenhouse effect on sea level rise have forced some reassessment of reclamation.

In the South Tarawa urban management plan several key land issues were identified that appear on similar lists for every country in the region (Kiribati Ministry of Home Affairs and Rural Development 1995: 10). Land is an exceptionally scarce resource requiring good management and the production of more resources (via subdivisions and appropriate reclamation) for private and public sector interests. Land supply has been neglected by government with the loss of various financial, economic and social returns, alongside environmental costs through the misuse of land above water lenses (a particular phenomenon of coral atolls). Necessary changes to customary land procedures would reduce the powers of landowners with respect to those of the government. The Land Management Division of the Ministry of Home Affairs and Development has been responsible for two important new developments which are likely to transform land use planning in South Tarawa. These are the establishment of the Kiribati Land Information System (KLIS), comprising six land data registers, and an Urban Planning Policy and Procedures Manual (Kiribati Ministry of Home Affairs and Rural Development 2000a, 2000b). As elsewhere the critical issue will be their implementation.

Land development in Koror

The pressure to find land for development purposes on the narrow peninsula occupied by Koror, the capital of Palau, is unusually intense owing to increased tourism and a desire by hotel developers and other businesses to gain a presence close to the magnificent Rock Islands. Development pressures temporarily eased somewhat in 1996 because of the collapse of the KB Bridge linking Koror with the main island of Babeldaob, thus severing road connections with the airport and development projects such as a Babeldaob perimeter road and the new national capital under construction on the east coast. Since independence, Palau has become the target of numerous investment proposals from Korea, Japan and Taiwan in particular and land is an integral component. Tourism is seen as a major platform for national growth and economic development, but uncertainties over property rights are a serious barrier to both national and international investment in this and other sectors of the economy:

> Clear property rights are a fundamental requirement for a modern capitalist economy. If Palau seeks to pursue a path towards such an economy, issues related to land tenure and effective land use planning will need to be resolved as expeditiously as possible.
>
> (SAGRIC International 1996: 7-1)

As elsewhere, that poses problems.

Land reform is fundamental to Palauan economic development, but change involves critical potential costs; land reform may result in several unwanted social effects, such as the erosion of village culture, the possible emergence of a landless class and the uneven accumulation of wealth, whilst security of rights exists within the traditional system. As elsewhere in Oceania, the government has found land reform difficult, and rejected the provisions of the first draft master plan, formulated by external consultants, when it was presented to parliament in 1996 because of disagreement with the chapter on land (SAGRIC International 1996: 7-2), a reluctance to confront the power of traditional landowners, and unwillingness to move more firmly into the commercial arena.

Uncertainties over title and rights to land have reduced the pace of modern sector investment and development at a critical period of the country's early post-colonial history, though a slower pace of development has incidentally provided time to identify local priorities and the means of effectively controlling externally induced development. Palau's Compact of Free Association with the United States government was agreed much later than in the other former Trust Territory entities, resulting in a financial support package that will be in existence well into the second decade of the century. This financial safety net, alongside land constraints, provides an accidental opportunity to plan for longer-term development and avoid the large-scale alienation of village land for tourism that has already occurred extensively in its northern neighbours, Guam and the Northern Marianas (Saipan), at some cost to local culture, environmental management and access to services.

As in FSM, most land in Palau is 'public' land (70 per cent) and much of it is in the process of being transferred back to private ownership by the Palau Public Lands Authority. The remaining public land will be held by the sixteen State Public Land Authorities. Land registration has been underway for some time but is likely to take another 10 years to complete because of delays and a lack of resources in the Land Claims Hearing Office. Only Palauan citizens and corporations wholly owned by Palauans can gain title to land but foreigners are allowed leases up to 50 years. No taxes are levied on land under the Constitution and compulsory purchase, though possible, has been used just once, to acquire land for the International Airport.

The government has set up a land information system but slowness in registration and transfer means that this too will not have the coverage necessary to provide a basis for infrastructure planning and other forms of development for at least a decade. As in most other Pacific towns, there is a shortage of developable land in Koror, because of unavailability rather than any absolute lack. The strength of Palauan attachment to land was evident in the full-page press advertisement for a candidate (a High Chief) in the November 1996 primary elections for President of the Republic. Here, resolving land issues was regarded as one of four basic issues facing Palau, the others being the importance of cultural heritage, environment, and government reorganisation through a Constitutional Convention (*Pacific Daily News*, 19 June 1996). Once again, here as elsewhere, land issues have increasingly become political issues, as land values increase and development is slowed.

Addressing land priorities

Land is a ubiquitous, central and problematic feature of Pacific island development and is integral to urban development. The ability of governments to plan for and manage urban growth rests ultimately on their success in controlling land markets and dealing with tenurial matters; effective housing and investment policy requires an adequate supply of well located and serviced building land, and efforts to promote sustainable urban settlement will fail unless great care is taken to safeguard water catchments and minimise development on unsuitable areas such as swamps and unprotected coastal locations. The Yapese saying: 'The man is not chief, the land is chief' (quoted by Hezel 1985: 1) remains valid as Pacific island governments have invariably failed to mobilise land effectively for development or constrain inappropriate uses of it.

A substantial amount of the land in urban areas throughout the Pacific is held under customary tenure, largely beyond the reach of regulation and management, other than by its customary owners. It is therefore only exceptionally available for 'modern' development. The relative absence of freehold land has raised costs, as new developments are concentrated on expensive plots on sites that are not necessarily the most appropriate or contiguous; hence even basic land use planning is generally doomed to irrelevance. The present political systems suggest that this situation is unlikely to change in the near future. Consequently commissioned reports on key urban development issues in the Pacific rarely omit reference to the need to modernise and reform land management practices but few are prepared to indicate in their recommendations what might realistically be done.

A central issue in the debate about urban futures is the problem of addressing specific urban strategies of one kind or another when the fundamental issue of land is a basic (and missing) element of the discourse. Standard solutions to questions of urban expansion and the provision of appropriate amenities and services presuppose considerable public control over land development, a situation that rarely exists in the Pacific. Governments have thus pursued the possibilities of creating land by reclamation despite its enormous costs, even when there is no absolute land shortage. To rectify urban land shortages means exercising government powers, such as land expropriation, that are controversial and threaten the basis of traditional island societies.

Land problems, particularly those concerning tenure, cannot be isolated from their social and political context. Small Pacific countries recognise where priorities lie but have been unable to instigate registration, let alone changes that will inevitably result in a rapid shift in power and influence from the traditional to the modern. At the heart of the dilemma, many prominent power-brokers, either contemporary politicians or traditional leaders (and, in many places, these are the same people), own and control significant amounts of urban land. In Majuro, the former President simply ignored any notion of urban planning or public good to establish the commercial buildings of himself and family members in prominent positions on his own land in the town. Similar examples exist elsewhere. Even where towns have recently been constructed in otherwise unoccupied areas disputes and power struggles have arisen. In Ebeye the major landowners gave the Kwajalein Atoll Development Authority leasehold over more than 75 per cent of the island in 1989. This enabled the authority to develop plans to decentralise the population, establish commercial zones, expand recreational facilities, and build the island's first high school. However, subsequent disputes between the eighty traditional owners of the island over the distribution of lease payments created some uncertainty about the town's future. In small societies the interests of particular individuals and groups have considerable potential to dictate the structure of development.

Retaining traditional land practices is not necessarily detrimental to development. In some of the smaller countries, and the atoll states in particular, there are very few opportunities for modern commercial development in the urban areas, or anywhere else. Removing traditional tenure in the towns under these circumstances would involve major political and social disruptions in exchange for creating a more secure investment climate for foreign investment. At least in the case of Palau, the inability to transfer land quickly into private ownership may have incidentally enabled a more appropriate and sustainable form of urban and national development. However, urban development is not solely about commercial, private sector issues, and in the Pacific this is relatively less important. It also involves the effective provision of housing and other services and the control of deleterious uses. Accessible public land is extremely scarce, and usually only exists where it was excised in colonial times. In some countries, like the Marshall Islands, it is simply absent. Majuro has no public land use planning scheme, no provisions for recreational areas or parks and no public open spaces: the largest terrestrial passive or active recreational area is a basketball court. The absence of land reform – and thus land use planning – precludes efficient urban administration, and imposes considerable externalities on the public sector. These problems, as much as the wish for commercial development, indicate the need for more adequate land management.

The most difficult issue raised by the need for greater regulation of urban land is the manner in which this opposes the customary rights, and interests, of communities within the urban area, especially those of increasingly marginalised groups such as the Motu-Koitabu of Port Moresby. This is particularly challenging because rural populations, the majority of Pacific islanders and the kin of many urban residents, freely exercise a wide range of unchallenged rights over their land; and because most urban residents have long been accustomed to exercising complete control over land use and access, and have opposed public regulations. In Majuro, for example, urban residents have opposed regulations concerning solid waste and sewage disposal, earth moving, road construction and land use zoning that might restrict them in any way. (This has affected attempts to develop appropriate garbage collection and disposal policies.) Moreover, as the case of the Ahi Association in Lae shows, particular local community interests within towns can be quite different from those of other populations within the city. The practice of land management in the Pacific is highly privatised at a very small scale – within lineages and households – and thus conflicts with the requirements for urban land management at a very different scale. Paradoxically deregulation is non-existent in a climate of increasing privatisation. Even the shift to more individualistic ownership makes the control of land use no easier.

One of the more persuasive arguments in favour of the adoption of formalised private property rights is that this will 'ameliorate tendencies of individuals to seek private gain through public office or leadership status' (SAGRIC International 1996: 3-3). Whilst individual land ownership characterises wealthy countries, in the Pacific this might facilitate the emergence of a landless class, an outcome that is not culturally acceptable, though some of those living in squatter settlements are already in this situation. It is also questionable whether private property rights will do anything to lessen corruption by wealthy individuals, who are usually supported by outside development interests. Perhaps inevitably there is some movement towards formalising individual rights over customary land in and around the more developed urban areas, such as Apia and Suva (Ward and Kingdon 1995), where the pressures for change are greatest and the demands on land for public uses are most acute. Private acquisition is thus proceeding by stealth.

Throughout the Pacific the negative effects of poor land use planning are most apparent in the larger cities. A strong case can be made for the public acquisition of strategic land for essential infrastructure and other important uses, without the threat of continuing or excessive demands for compensation, but the three ways commonly used by governments to intervene in the use and ownership of private land – eminent domain, property taxes and zoning laws – are all highly unpopular (Hezel 1994), and have thus been little used. Pacific islanders are no different in this respect from residents of cities elsewhere in their suspicion that inadequate compensation will be forthcoming, and that those who cede their land rights will be marginalised and are consequently the most ardent advocates of 'not-in-my-backyard'.

Practical progress can be made in urban land reform, as new managerial initiatives in Kiribati may demonstrate. Public acceptance might be gained in the largest towns for limited use of eminent domain powers to acquire land for essential community purposes such as the improvement of urban sanitation services. This will never be easy; though the legislation is already in place in the majority of countries, the weakness of urban local government is everywhere a constraint and places much of the responsibility for raising public awareness on national government and NGOs. Greater priority and more resources are slowly being allocated to the process of urban land registration to ensure that extensive backlogs in survey work and the granting of titles are cleared, yet the sheer burden of the work, the cost and scarce skilled resources have slowed most efforts. Limited successes of this kind provide a catalyst to identifying other reforms. The tasks remain enormous:

> Land management in the South Pacific . . . requires considerable professional skills. It requires a good knowledge of the relevant traditional cultures, the received legal cultures and the relationship between them. It requires some knowledge of practical skills such as surveying, valuation and accounting, if only to give general guidance and to know when to call in specialists. It requires skilled record keeping.
>
> (Ward 1994: 48–9)

Even the best land reforms will fail without the presence of this range of skills. Yet the limited availability (and commitment) of appropriately skilled labour, poor coordination between departments, inadequate finance and a lack of clear objectives – exactly the constraints that hamper urban management generally – are even more evident in land departments. Despite the growing pressure on urban land throughout the region, the crucial need for management (especially where environmental deterioration is occurring), and the constraints that inadequate land supply pose for social and economic development, it has never been given any degree of priority anywhere in the region: a powerful indication of the continuity of Pacific island cultures. Land questions are unlikely to be quickly or painlessly resolved as long as land and identity are intertwined, yet they are at the core of sustainable urban and national development.

7 Housing and the urban environment

> Mele who helped in our house came from Vava'u to give her seven girls a better chance. She lived in a packing case in a swamp in Sopu. Her sister had a small concrete house nearby. When it rained and the water rose too high, Mele et al moved to her sister's house. She bought coral rock to raise the land level and tried in vain to get some rights to the land. An area of land was allocated by government for low-cost housing after the cyclone. It was very swampy and aid organisations said it was no good. Mele bribed her local noble but to no avail. The land was later taken over for high-cost housing. Bad luck, poor people. They remained where they were in their rubbish bag and scrap village.
>
> (*Te Amokura*, January 1994)

Housing in Pacific towns – a good indicator of other features of the urban condition – varies enormously and underlines the uneven and diverse experience of urbanisation. The proliferation of squatter areas in Melanesian towns, for example, is less evident in Polynesia and Micronesia though there are pockets of informal and overcrowded urban housing there (evident in the above example from the Tongan capital of Nuku'alofa). Dwelling densities in Ebeye and Funafuti atolls, which are among the highest in the world, contrast with much larger plot sizes found in Koror or Apia. Many houses in South Tarawa are comparable to village dwellings, with few services of any kind; in PNG's mining towns most houses are formal, modern and with excellent services. There are also similarities. Urbanisation pressures throughout the region have made it increasingly difficult for governments to address housing needs and demands, regardless of national circumstances, and dealing with the lack of affordable housing for low-income residents (who make up the majority of the urban population of the Pacific) and ensuring basic service provision are some of the most intractable and visible elements of the urban scene.

The quality of urban life in the Pacific involves several key characteristics. Housing is the most basic of these. Others include the availability of employment, access to health, education, water (and other services such as sewerage, waste disposal and electricity) and the state of the physical environment. Though Pacific towns rarely experience the extremes of industrial pollution found in some larger Third World cities, deterioration in living conditions is occurring because of the poor state of public utilities throughout the region. Increasing demands on limited water supplies, together with pollution from uncontrolled domestic wastes, for example, threaten the sustainability of present levels of urbanisation in the Pacific and will ultimately affect the success or even relevance of government housing policy. Examining the role of housing as a development issue usually involves an artificial separation of concerns about physical provision and maintenance of shelter from other associated aspects of the housing process such as land supply and opportunities for rural and urban employment. The most important priorities for governments to address when seeking improvements in urban housing policy may be far removed from considerations about the condition and production of dwellings themselves but lie in issues like land supply, housing affordability and provision of infrastructure. This chapter examines the diversity of housing conditions in the region, the relationship between housing and other services – primarily water and waste disposal – and the development of policies in these sectors.

Housing provision

Several interrelated factors play an important role in the success of government housing strategies in Pacific towns. The colonial legacy is of significance almost everywhere; there are marked differences in building costs and development opportunities caused by locational factors; variations in national wealth and other resources influence the provision of urban amenities; and specific socio-economic and demographic features affect the character of national and local housing markets. Building costs are lower in the larger states where some local resources are available, whilst the more impoverished countries are least able to put in place effective housing programmes. Differences between migrant groups (especially in the larger Melanesian towns) and the strength of links with rural origins complicate both housing demands and socio-economic wellbeing. Polynesian towns are far more homogeneous and a considerable amount of housing stock may be financed by remittances from relatives working overseas. The contrasting effects of such socio-economic and demographic factors ensure a diversity of housing needs and varied demands on housing stock and services.

Severe population pressures of the kind experienced by the Melanesian capitals in the 1970s and 1980s and in Micronesia in the 1970s were fed by high rates of rural–urban migration causing urban housing problems that governments could not deal with effectively. An influx of rural poor had little option other than to settle on customary or government lands in and around urban areas. Sufficient basic infrastructure was not provided to these areas in advance of settlement nor could enough public rental accommodation be built. In post-colonial times especially, the maintenance of existing public housing deteriorated, and the supply of adequate residential land was never sufficient (Chapter 6). By contrast, outmigration from the principal Polynesian cities

Plate 7.1 Sopu, Nuku'alofa, Tonga. Housing has extended into extremely flood-prone areas on the edges of the lagoon. © John Connell

during the same period relieved urban growth pressures considerably, resulting in housing problems of a different and less intense kind, though land supply remained a challenge (evident from the quotation which opened this chapter). Environmental factors and limited land areas in the atoll states have greatly curtailed opportunities for urban expansion of a sustainable kind because of the small capacity of natural sinks to cope with waste disposal. Housing policy responses by Pacific island governments have been affected by several dilemmas. Beyond the perennial financial problems of providing affordable housing for a low income population, there has long been some aversion to urban policy formation. The success of residential upgrading policies makes towns more attractive destinations for those living on outer islands and remote rural areas where little in the way of a modern life style is available. But failure to turn Pacific towns into better places to live, because it would make them the target for even more unmanageable growth, is neither a sufficient reason for governments to shelve the important goal of urban improvement, nor should it be a politically feasible strategy.

Social factors, such as the changing character of the extended family in the Pacific, contribute to the nature of the housing process. Thus since 1950 in Pohnpei (FSM), a seemingly irreversible movement towards nucleation of the household unit has led to greater demands for housing that cannot be explained by crude population growth alone (Barnabas and Hezel 1994). However, elsewhere in Micronesia, particularly in South Tarawa, a preference for the continuation of extended household living is evident and quite striking (Atoll Research Programme 1996) whilst in several cities urban household sizes have continued to grow (Monsell-Davis 2000). Nevertheless the gradual emergence of more nuclear families and individualistic life styles alters the rate of household formation and the utility of the dwelling itself. Economic factors, including incomes and remittances, affect the housing situation. It would be difficult otherwise to explain the existence of substantial modern residences in several Polynesian and Micronesian towns whose high costs are unrelated to local income earning opportunities.

Housing contexts vary among countries and towns. The small amount of formal housing, either present (or former) government or private dwellings, is actually outnumbered by an extensive informal housing sector in the largest Melanesian towns. This large collection of self-built dwellings is made up of traditional villages swallowed up in the expanding towns and enclaves where migrants and settlers (a PNG term) have squatted. Squatter settlements make up a significant proportion of all Melanesian urban housing areas. By 1986 about one in eight persons living in Suva was a squatter, and that proportion remained about the same in 2000, although one official estimate stated that more than a quarter of Suva households were living in informal settlements and fringe villages (Government of Fiji and UNDP 1997). In Port Moresby as many as half the population may be housed in settlements (Koczberski *et al.* 2001). The term 'squatter', in the Pacific, refers to those living in more or less unplanned, informal settlements, most of whom have permission to occupy land and pay rent (in contrast to traditional landowning groups, who are not squatters but whose often informal dwellings resemble those of squatters). In Fiji more than half the squatters (over 56 per cent) were Indo-Fijians and almost 40 per cent were indigenous Fijians (Bryant 1993) but most squatter settlements there and elsewhere exhibit considerable ethnic and social diversity. Dozens of distinct settlements characterise cities like Port Moresby and Suva, whilst even Alotau has as many as eighteen. Blacksands in Port Vila and the largest PNG settlements have more than 5,000 residents and possess a range of facilities. Many are located on marginal land, sometimes in low-lying areas, where waterlogging and inadequate sanitation systems pose health hazards. In earlier years, around the time of independence, Pacific governments sought to provide assistance to settlements through infrastructure provision, as at Ranuguri, Port Moresby (Forbes and Jackson 1975), but by the 1980s and 1990s settlements were receiving few resources, and squatters had become the pariahs of urban development to be opposed rather than assisted. Nevertheless most settlements exhibit income and employment structures quite similar to those of the urban area as a whole, with both skilled and unskilled workers, and considerable permanency (Connell and Lea 1993: 119–20; Goddard 1998; Mecartney 2001). Settlements are not necessarily the first place of urban residence. At Blacksands many residents had made several moves within the city before finishing there, some through a clear preference for Blacksands – for greater security of tenure, additional space or closeness to kin – and others, in

Plate 7.2 Morata settlement, Port Moresby, Papua New Guinea. © John Connell

Plate 7.3 Three Mile settlement, Port Moresby, Papua New Guinea. © John Connell

recent years, because of the destruction of nearby houses by Cyclone Dani (1998) and the 'right-sizing' (less euphemistically 'down-sizing') of the public service through the Comprehensive Reform Programme (CRP) which created new unemployment (Mecartney 2001: 76). In every settlement housing improvements have occurred. More than a third of Blacksands households lived in permanent houses, with concrete foundations, finished timber construction and galvanised iron roofing, though most houses were sparsely furnished with earth floors and separate outside 'bush' kitchens (Mecartney 2001). Even relatively poor settlers, with insecure tenure, have significantly improved their residential status.

In most of Polynesia and Micronesia by contrast (other than in Kiribati and the Marshall Islands), informal sector housing is not nearly so extensive, and in some towns, like Apia, Kolonia and Nuku'alofa, it resembles collections of urban villages. Old suburbs in the faster growing centres are rapidly losing inherited colonial characteristics of bungalows and flower gardens (usually alien to indigenous people) in favour of something more typical of the modern Pacific. The change in appearance accompanying urban growth, however, is sometimes unwelcome where management is lacking:

> Here is a small town battling a terminal illness threatening to put her down. Filth and disease are swimming in swamps and puddles everywhere. . . . The saddest thing of all is that nobody seems to know what to do. Where are the government's so-called thinkers, the town planners and engineers? Why hasn't anybody been able to figure out how to move the disease-ridden swamps of Fugalei and Taufusi into the sea? The truth is that Apia's growth has been advancing rapidly over the last few years, so that now it is far ahead of development. . . . Even though it is plain as daylight that proper drainage and sewage systems are as essential as food and water, they have been taking the lackadaisical attitude. . . . Apia in its present state is a filthy, depressing little town, totally empty of inspiration. Its ugliness constantly defies the imagination.
>
> (*Samoa Observer*, 25 February 1992)

Most Pacific island countries have made considerable progress since independence in identifying housing problems and determining the most appropriate mix of policies to resolve them. In this chapter the background factors identified earlier are related to government experiences in developing effective urban housing policy.

Overcoming historical legacies in Melanesia

Only recently has urban housing in Melanesia moved away from the overwhelming domination exercised by employer-provided accommodation in the colonial period when many jobs were in government, the private sector provided tied accommodation, and living in town was considered a temporary phenomenon for indigenous people. Domestic employees in PNG were even accommodated in a *boi haus*, many of which survive in the grounds of large houses. The situation where no provision was made for Melanesian family housing and wages were adequate only for single men (Chapter 2) persisted into the post-colonial era until informal settlement became the prevalent pattern of urban residential development. Pools of highly subsidised and high-cost houses were created in the Melanesian capitals, mostly owned by the government, together with a small number of formal sector houses for purchase or rent on the open market. Few of these were (and are) affordable, even for middle-income households, and remained quite outside the reach of most urban residents. Over the years Melanesian governments sought to reduce their direct involvement in housing through such schemes as the construction of low-cost homes for sale, rental of government houses, support for resettlement schemes and provision of subsidised finance for intending house purchase. However, the decline in the proportionate significance of government housing throughout the region has occurred because of the huge growth of informal settlements where public intervention has been minimal.

Increased demand for formal housing in recent years has been a consequence of low average family incomes in most urban areas, limited finance for housing loans (especially for low-income families), a concentration of financial benefits on those with ability to pay, high building materials and construction costs, an undeveloped low-cost housing market, and construction bottlenecks caused by the limited pool of tradesmen with technical expertise. These problems have been accentuated by high rates of rural–urban migration and natural increase, the small number of private sector housing construction companies, poor administration and management of public housing programmes, and public housing subsidies of one kind or another to those who are less appropriate for such help than the relatively poor.

The situation in individual Melanesian countries differs considerably in the scale of housing operations and the characteristics of prospective tenants and purchasers. Generally, however, the poorest sections of the urban population only receive effective assistance from relatives, non-governmental organisations (NGOs) or the churches, emphasising the gulf between official policy and actual need. Moreover, public housing authorities have found it almost impossible to manage rental housing stock successfully. In Fiji, for example, the Housing Authority adopted a three-part strategy in the 1990s: to lease good quality land in Suva at above development costs to the relatively wealthy; to generate enough revenue to provide housing at cost for those purchasing its houses; and to supply land at below cost for sitting tenants required to make way for more intensive development. This policy reorientation followed the Authority's increasing difficulty in maintaining, paying for and adding to the stock of its rental housing. Other countries have found this even more difficult than Fiji. Housing authorities were usually established towards the end of the colonial period as welfare agencies to provide subsidised accommodation for the urban poor. Emphasis on welfare through rent rebates gave way to a 'user pays' approach in the 1980s as deficits mounted and rental arrears increased, whilst the extent of informal housing increased.

The housing situation in Fiji in the 1970s was characterised by numerous problems, including the persistence of high costs and the presence of regulations demanding high dwelling standards, which resulted in some 80 per cent of public housing applicants in Fiji not being able to afford the cheapest houses (Walsh 1979, 1984). Since then access has become more difficult for those with little chance of receiving government assistance, forcing them to live at the urban fringe and in new squatter settlements, though squatters in old-established sites were in a better position to take advantage of available government housing schemes. Restructuring in Fiji has made little difference. In 1989 the Fiji Housing Authority was separated from a newly created Public Rental Board, which was left with a large loan liability and responsibility to manage the remaining housing stock at economic rentals. In the 1990s the Authority concentrated on developing new subdivisions and only selling houses to those in the top half of the national income distribution. As many as two-thirds of Public Rental Board tenants in 1992 were long-term residents of subsidised flats and could afford rents in the private sector but failed to relocate, weakening the ability of the Board to target its accommodation effectively (Government of Fiji and UNDP 1997). Few of the poor could afford housing aimed at those earning between $F45 to $F160 a week, when the average monthly minimum urban wage in the early 1990s was $F45 per week (Bryant 1983: 49). Government institutions have failed to cope with the demand for affordable shelter by providing cheap accommodation.

This inability of government or the private sector to provide affordable urban housing in Fiji placed great importance on the roles of NGOs and churches. The Housing and Relief Trust (HART), a church-based organisation, assisted some 1,570 low-income people between 1971 and 1985 (Bryant 1993). Plans by the Methodist Church in 1992 to purchase most of a large squatter camp in Suva for sale to its occupants drew media comment: 'It is a pity that it is the Church that has to shoulder a burden that rightfully belonged to government' (*Fiji Times*, 19 November 1992). By 1994 the Methodist Church was running poverty alleviation projects in three Suva squatter areas, involving more than 700 low-income families at a cost of some $F1.5 million (Government of Fiji and UNDP 1997). The importance of NGO and church assistance in improving the housing and living conditions of the poor in Fiji was recognised in increased government funding for HART and plans for joint government and NGO upgrading projects in squatter areas:

> Meeting the housing needs of the poor requires the support of all stakeholders: the NLTB, the Housing Authority, Lands Department, PWD [Public Works Department], city and municipal authorities, and NGOs, such as Rotary and HART, that provide low-cost housing. . . . Efforts should be made to enable the poor to participate in finding solutions to their problems.
>
> (Government of Fiji and UNDP 1997: 117)

The provision of low-income housing has become such a complex and difficult question, even in one of the more developed countries, that it demands action by a host of public and voluntary organisations and the residents themselves, but in a vague and unspecified manner. Private sector initiatives do not extend beyond the upper income reaches of the housing market.

Elsewhere in Melanesia the housing situation has evolved similarly. In Vanuatu provision of low-cost housing before and after independence was frustrated by the peculiarities of the colony's legal restrictions and, by the early 1970s, 'almost nothing had been achieved' (Bedford 1973: 98–9). The National Housing Corporation, established in 1984 as a typical housing rental agency with its priority being the needs of the poor, subsequently revised its objectives up-market to 'produce and encourage the production by others of shelter for families who do not otherwise have available to them the kinds of living accommodation they would both desire and afford' (Asian Development Bank 1997: 58). This followed a succession of project failures and the particular difficulty of providing accommodation for the urban poor. The first housing project (not completed until 1988) collapsed 'through mismanagement and failure to provide utility services into the low cost housing subdivision . . . only ten out of a planned forty houses were sold' (ibid.: 57). The Corporation was substantially undercapitalised in 1997, its only assets being 400 urban serviced blocks in Port Vila and fifty in Luganville, and it had no effective Board, a situation that reflected the inadequate financial resources of many urban agencies compounded with the lack of managerial skills, and provided no solution to worsening housing problems.

A recurrent issue dating back to the self-help debates that preoccupied the developing world in the 1970s, which complicates all Melanesian urban housing provision, is that of high public housing standards and their relevance to the needs of the urban poor (Lea 1979). Most housing authorities in Melanesia that still provide public housing have reduced construction and amenity standards to more affordable levels. In many instances this has meant providing no more than access to a serviced block where tenant-financed self-build housing can take place: the legacy of an earlier era. In the Solomon Islands the need to relax certain standards to contain costs has never become evident at a political level. In the low-cost Honiara housing estate of Naha Valley in 1983, the builder of a 120-house project (funded through Australian foreign aid) was criticised by the then Prime Minister for building 'rubbish', though the project had been approved by the Public Works Department under conditional zoning for low-cost housing, under which normal statutory standards were waived (Sofield 1993). The shift towards relaxation of standards also followed attempts to incorporate elements of local design for cultural reasons and because local housing demonstrated the resilience and utility of traditional building methods to cope in cyclones. Despite its cost and design advantages, Solomon Islands traditional housing was not permitted under the building by-laws in the Honiara town area until recently, whilst resistance to appropriate housing has been widespread. Public welfare objectives exercised through housing assistance have generally failed in Melanesian towns, hence the need for innovative and integrated schemes which only Fiji seems close to achieving.

Most formal housing in Melanesian towns consists of detached dwellings at very low densities, of the kind found commonly in parts of northern Australia, with considerable additional costs in providing adequate transport, sanitation and other infrastructure. High-rise apartments and other forms of urban consolidation, notably town houses, are belatedly becoming more common in some of the capitals, especially in Port Moresby where security issues are serious. Public housing in PNG was mainly concentrated in a few inner urban areas of Port Moresby in the early 1990s and was supplied by government to senior national and expatriate employees, with the majority of houses eventually being sold or given away to the tenants. The cost of living in rental accommodation is high in all PNG urban centres primarily because of its very limited availability. A broadly similar situation exists in the Solomon Islands and Vanuatu, though neither country had much high-cost formal housing. Heightened housing costs led to public service salaries in Melanesia being extremely high in regional terms and contributed to raising the costs of government. Alongside high wages in the private sector, partly necessitated by housing costs, this has discouraged the establishment of manufacturing in PNG cities.

Demand for housing has been much greater than supply, though few studies of preferences have ever been undertaken. At the end of the 1970s in PNG there were reported to be over 5,000 applicants waiting for National Housing Commission houses (To Mutnaram 1979: 2) but this represented only a fraction of real demand, despite the fact that only four out of ten people living in towns at that time could afford to rent the cheapest Housing Commission house available (Jackson 1978: 172; Norwood 1979, 1983). Similarly, in Honiara in the late 1980s, 58 per cent of the permanent urban population could not afford even the cheapest house. If land costs were added, this percentage was even higher (Honiara Town Council 1988: 33; cf. Nage 1987). In Fiji much the same was true: in 1976 some 8,207 households, or 22 per cent of the total number of urban households, were on the waiting list of the housing authority (Chandra 1979: 29). Surprisingly, the extent to which these situations have changed is still largely unknown and there are remarkably few data on housing preferences and affordability. It is a measure of the limited focus on urban development issues and the inability of housing authorities to respond and cope that such basic information is unavailable. Nevertheless demand continues to far exceed supply, and access to the small amounts of national housing stock available is necessarily limited. However straightforward and egalitarian application procedures may appear in theory, access to housing tends to be inequitable in practice; in PNG 'an elite is therefore established. It may be an elite of those who can speak English, or it may be an elite which happens to have *wantoks* [kin] in the Housing Commission. Whichever it is it makes meaningless the idea of queueing for access to housing' (Bryant 1977: 50). Queueing, in any sense, has become even more pointless as demand has intensified.

Provision of urban housing throughout Melanesia has taken two broad forms: promoting individual self-reliance through the supply of subsidised plots, enabling residents to construct their accommodation in their own way; and assistance in securing commercial bank loans for private financing of land or house purchase (Asian Development Bank 1992: 37). The first form of provision, by means of 'site and service' schemes, was common in PNG and became significant for encouraging self-built housing by low- (or non-) income earners, many of whom were recent migrants to the city. The second method was more relevant for the needs of relatively high income earners with permanent jobs. (A third form, the construction of government housing for sale or rent to medium- and low-income households, still persists in a minor way, though provision of public rental accommodation is now relatively unusual because of problems encountered by housing authorities with large-scale rent arrears and high maintenance costs, and their own inefficiency.) Housing the high-income sector is catered for by private building companies but political uncertainties in several countries, such as the instability experienced in the wake of coups in Fiji in the late 1980s, have acted to reduce construction activity considerably, raising the cost of existing housing stock and imposing further costs on urban economic activity.

The high cost of formal rental housing has ensured that there is very little anywhere in Melanesia. In PNG, 'even a highly paid expatriate employee, earning twice the salary of a Papua New Guinean counterpart, would not be able to afford to pay the market rent for a medium-cost three bedroom house' (Dahanayake 1991: 2). Most of those on middle incomes in the private sector live in employer-provided housing with highly subsidised rents. Such employment conditions for salaried employees are considered essential in PNG, where rentals in the large towns can be double the average of those in such metropolitan capitals as Sydney or Auckland. High rents are partly the result of a low rate of formal housing supply in a country where the number of houses constructed during 1980–9 was less than 10,500, whereas urban households were estimated to have increased by 36,700 over the same period (ibid.: 9). Effective demand has far exceeded supply in other parts of Melanesia, resulting in a broadly similar situation.

In the mid-1980s the PNG government chose to move away from major involvement in the housing market, and sought to rationalise housing policy by facilitating a free market in housing but, after political and trade union pressure, the new policies were weakened and only the sale of government houses was actually implemented (ibid.: 14–16). A new housing policy package was adopted in 1989 which again sought considerable government involvement through a large housing bureaucracy (including the creation of the National Housing Corporation (NHC)) and substantial subsidies; it also intensified bureaucratic control and regulation rather than enabling a gradual shift towards the market economy (ibid.: 17–18). Housing policy in PNG thus tended to favour large-scale interventions through subsidies rather than the removal of causes of market failure. This too was echoed in the policies adopted in the Solomon Islands and Vanuatu.

The sale of high-cost public housing in PNG was first debated in the 1970s when it was argued that this would result in greater environmental and community awareness because of owners' committed stake in the maintenance of property values (Curtain and Blaxter 1978). Home ownership was promoted by stressing the concept of the house being a valuable and transferable commodity (Stretton 1979). But deregulation proved difficult throughout the region. When the Solomon Islands government first gave its public servants an opportunity to buy the houses they occupied, the Civil Service Association opposed the move, resulting in the abandonment of the policy when a housing shortage was recognised (Nage 1987: 96–7). None the less, the Solomon Islands government subsequently returned to a policy of selling its houses, like the other Melanesian states.

In 1992 the PNG government decided on the extreme solution of a 'Give Away Scheme' that relinquished ownership of all lower-cost public sector houses to their existing tenants at no cost, to encourage home ownership and relieve the public sector of responsibility for maintenance and management of the properties. Some 6,000 houses valued at K24 million were involved. The scheme was initiated in the Port Moresby suburb of Gerehu in 1993 and subsequently extended to other suburbs and towns. The decision followed difficulties experienced by the NHC in administering the properties and the refusal of many tenants to pay the rent because, they maintained, of the physical deterioration of the houses. Rental arrears at the beginning of 1992 had risen to over K2.5 million (*PC*, 30 September 1992) and were effectively written off. Even deregulation at this scale had only a small impact on the overall housing system. It was an acute measure of the various failures of public housing provision.

Perhaps the most serious constraint on urban housing provision in Melanesia is the limited availability of building land and the failure of government policies to recognise this. In PNG the provision of urban housing land was never confronted until the Minister of Housing introduced a housing policy statement in 1992, noting that enough land would be subdivided and made available for people to build their own homes. The Land Mobilisation Division of the NHC subsequently experienced problems in arranging sufficient land supplies (Papua New Guinea. NHC 1992: 10–11) and it was never explained where the land would come from or how it would be made available for residential development.

A second crucial urban housing constraint is availability of finance under circumstances where strict limits to mortgage funds have proved a major barrier. The Papua New Guinea NHC introduced several new housing finance schemes in 1992 based on a Housing Refinancing Facility (HRF) aimed at first home buyers with annual incomes less than K15,000. The HRF had five schemes, each designed for particular groups of potential homeowners. The Grassroots Toea-Savers Housing Scheme (GTHS), for example, targeted the lowest-income households earning less than K3,000 (about two-thirds of the urban population in the early 1990s) who were not regularly employed or were unemployed: an innovation since all previous schemes had demanded that participants be regularly employed. Those interested in the scheme were expected to demonstrate an ability to save by putting aside a sum of about K2,000 for a period up to two years that would enable them to raise the amount appropriate for a deposit. Although the NHC identified two particular groups who comprised the most visible elements of the informal sector found in most PNG towns, the bottle collectors and betel-nut sellers, it had never determined either their patterns of income and expenditure (and thus ability or propensity to save) or their aspirations towards formal home ownership. In other words, the policy had been designed without regard to the needs, aspirations or ability to pay of the intended recipients.

The conjuncture of unemployment and poverty (or inadequate incomes from informal employment) in the larger towns suggests that housing policies should give greater focus to low-cost or self-help housing, because of the high proportion of settlers. Moreover, as in PNG, 'the dominance of large foreign building firms, the use of imported building materials and high building standards have discouraged the development of a low-capital building industry' (Asian Development Bank 1992: 53). A further constraint to the development of national building industries is the difficulty of identifying and obtaining suitable plots of alienated land. Although there is a general policy bias towards private enterprise and free market conditions in Melanesian housing markets, many constraints prevent them from operating efficiently. The more obvious include the lack of an open land market and problems in securing development finance, resulting in shortages and exorbitant prices, especially in the capital cities.

Rents and payments for other housing services have been difficult to collect in Melanesian countries, even though most formal housing is owned or occupied by an elite of high-income earners. At the end of the 1970s, for example, housing authorities in PNG and Fiji had accumulated significant rental arrears, such that they virtually stopped providing new rental accommodation for low-income tenants in Port Moresby and Suva. A new policy of subsidising the construction of houses for sale was introduced. In Suva, the change scarcely solved the problem because the Housing Authority was still responsible for collection of mortgage payments on dwellings sold to the public. Collection efficiency was found to be no more than 60 per cent in 1991 and some mortgages were two years in arrears (UNDP and UN Centre for Human Settlements 1991; Bryant 1992). The financial and administrative problems of government housing agencies were not simply confined to coping with the housing needs of the very poor in the settlements but also extended to the formal housing sector, making adequate management even more difficult to achieve.

The housing difficulties of Melanesian cities in the post-independence period are largely outcomes of processes commonly found in rapidly growing centres. Many migrants house themselves on land owned by other groups, whilst the small wage- and salary-earning elite has been forced to accept some of the burden of acquiring modern dwellings by governments no longer prepared to pay for the subsidies introduced by previous colonial regimes. Less prominence has been given to housing as a product, because the physical side of dwelling provision is usually given a lesser order of priority in Melanesia than associated concerns of land supply, basic infrastructure and sanitation services.

Housing stress in the atoll states

In the atoll states of Kiribati and the Marshall Islands, the issues are different since the limited physical environment has forced governments to confront and manage some of the highest housing densities in the world. Effective urban administration, efficient use of resources and good planning are thus at a premium. The speed and intensity with which urbanisation took place in the Marshall Islands in the 1960s was related to US expenditure on its missile base in Kwajalein, changes in the administration of the Trust Territory, rising expectations and better transport and communications. Between 1963 and 1970 the populations of Majuro and Ebeye grew to over half the national total for the Marshall Islands as a whole. The new towns soon reflected their rapid and haphazard beginnings: 'a melange of a few modern concrete structures and dilapidated relics from the US naval period, permanent homes and shabby houses hastily constructed from any material at hand to house immigrants from other islands . . . haphazardly wherever land was available' (Hezel 1995: 324). The American administration had no obvious interest in developing an appropriate housing policy anywhere in Micronesia, despite its funding of elaborate town plans. A very different colonial heritage, and steady rather than rapid growth, have resulted in contemporary housing scenes in Kiribati and Tuvalu quite different from each other, and from the Marshall Islands, despite similar environmental conditions.

Marshall Islands

The most extreme urban pressures developed on Ebeye island, home to islanders working on Kwajalein missile range or relocated from islands in the missile impact zone. Initially the living conditions were scandalous (Connell 1983c), but by the 1970s a US$5 million improvement programme had led to better housing and marginally improved environmental conditions with the construction of 300 living units incorporating some modern amenities, provision of communal toilets, a power plant and sewerage system. But Ebeye remained 'an island of affluent paupers' (Hezel 1995: 327) and even the expenditure of many millions on sanitation services, desalination and other infrastructure was said to have led only to a healthier kind of slum:

> Everywhere there were the carcasses of cars, split bags of household waste, beer cans, disposable nappies. . . . Almost treeless, sanitationless, overrun by children [yet] the sewers no longer backed up into the one-room houses.

> They had desalination and power plants. The streets and sidewalks were paved. Hepatitis was no longer endemic.
>
> (Evans 1992: 262–3)

Establishment of the Kwajalein Atoll Local Government (KALGOV) introduced urban administration to Ebeye and a Kwajalein Atoll Development Plan guided infrastructure development, but residential densities at over 28,000 persons per square kilometre in the 1990s had reached saturation levels. Housing as crowded as this is only sustainable as long as the continuing presence of the missile range generates wage employment and sufficient subsidies to pay for expensive infrastructure.

There is an effective policy vacuum in the Marshall Islands. The national housing authority was dissolved in 1991, and responsibility shifted to the tiny Housing and Home Improvement Division of the Ministry of Social Services. In effect this devolved housing activity to the private sector – a high-cost sector, since most materials are imported from Hawaii, with little relevance to low-income populations. As in Palau and the FSM there is an absolute lack of appropriate housing policies, a legacy of colonial policy, hence one critical element of urban planning is absent.

Kiribati

In the other Micronesian atoll state of Kiribati the government completed the first national shelter strategy in the region (Kiribati Housing Corporation 1997), after commissioning detailed studies (Atoll Research Programme 1996; Kiribati Housing Corporation 1996). This interest is not matched elsewhere in Micronesia, where there is a widespread perception that housing provision is primarily a private activity to be undertaken by households rather than government. Throughout former 'American' Micronesia this explains the surprising lack of emphasis on most aspects of urban development and the prominent role of NGOs and churches. Although Kiribati has physical and cultural similarities to the Marshall Islands, differences in its colonial history and economy have resulted in a quite different urban administrative environment. The British colonial administration set up parliamentary and local government institutions before independence and a process of public intervention in some aspects of urban affairs was established; yet in a remote, impoverished colony substantial financial support for infrastructure development has never been present, hence urban services are relatively undeveloped.

By the 1990s urban pressures in South Tarawa were alarming, despite efforts at decentralisation, and a housing study was undertaken in 1995 because of the need to address the steadily deteriorating situation. In the most populated urban island of Betio the density had reached 7,000 per square kilometre and densities in other critical areas were almost as high. Numerous problems had become readily apparent, including the lack of access to housing and land; the lack of access to clean water, sanitation and power; the presence of informal housing

using poor-quality salvaged materials; the increase in the incidence of squatting; high occupational densities, causing a lack of privacy and disorganised and unplanned building; the nuisance caused through noise and smell; and the loss of visual amenity (Kiribati Housing Corporation 1996: vii).

The chief factors underlying the housing crisis in South Tarawa are not all capable of remedial treatment through housing policies alone since the inadequate urban shelter situation reflects broader social and economic issues. Whilst Kiribati has sought, in the 1997 strategy, to deal with access to land, provide clean water and sanitation, develop and enforce building standards and obtain funds to underwrite maintenance programmes (Kiribati Housing Corporation 1997), even in a context of a planning programme the ability to achieve effective change has been constrained by inadequate financial and human resources and political wills. Comprehensive housing strategies require administrative support.

Tuvalu

In the smallest atoll state, housing problems have been quite recent, following secession from Kiribati and the rapid growth of Funafuti. Almost half the population (3,850 in 1992) of Tuvalu live on two islets of Funafuti atoll in some 500 private houses at a 1991 density of more than sixteen persons per hectare (1,600 per square kilometre). Fongafale islet, the seat of national government, had a population density of over twenty per hectare in the early 1990s, a figure projected to increase to forty-seven persons per hectare by 2006 (Clarke 1993a: 12). Densities of this magnitude, coupled with small, finite and fragile land resources, raised the spectre of a 'doomsday scenario' for the nation (Cole 1993), resulted in domestic pressures for emigration and indicated the need for comprehensive measures to ensure sustainable development.

Housing issues in Funafuti were of concern as early as 1981, when overcrowded housing was a subject of parliamentary debate. It was also debated whether it was appropriate to provide housing for all government employees and whether government workers should be appointed to Funafuti from outer islands, given the responsibility to house them that this entailed. In part the crisis reflected a shortage of housing thatch (and firewood) and greater dependence on imported materials in the more 'traditional' housing sector. Most contemporary houses in Funafuti are permanent structures with cement floors, aluminium roofs and electricity; less than 20 per cent are of traditional materials with thatched roofs. Most public servants are accommodated in government housing, maintained by the Public Works Department, and there is virtually no private housing market. No government ministry is responsible for overall housing policy, hence little attention has been given to preparing a national housing development strategy. Limited land is available for new construction other than on reclaimed areas where costs are considerable, a situation exacerbated by continued rural–urban migration and problems that followed separation from Kiribati, when it was agreed that public servants could own their homes after 10 years. It was then believed that public servants would wish to retire to their home islands, but few have chosen to do so. As the Prime Minister has noted: 'nobody wants to go back home. They all want to retain their houses. The landowners are up in arms, saying you [the government] are not going to acquire any more land' (quoted in *Pacific Report*, 4 April 1994: 3). This has led to overcrowded housing and considerable pressure on land, water and other resources.

Housing development has largely been a response to particular needs. A succession of housing initiatives followed the comprehensive destruction of all dwellings on Funafuti Atoll by Cyclone Bebe in October 1972. Fifty-five public service houses were requested of aid donors in 1984–5 as the government bureaucracy grew, but after design problems and extensive cost overruns, a 1990 appraisal mission recommended closure of the project. The mission identified several key housing issues: the assessment of housing demand including standards and affordability for the public and private sectors on Funafuti and the outer islands; the development of an equitable method of delivery of housing including both private and public finance; and the need for a comprehensive land planning management policy. The eventual upshot was the commissioning of a comprehensive housing sector review (Clarke 1993b). Like its former colonial partner, Kiribati, Tuvalu was moving into a new era of housing and development planning. Yet while the housing review recommended comprehensive changes to land registration and management (Connell and Lea 1995), that were typical of strategies proposed elsewhere, they depended for their adoption on major change in attitudes towards land and development, and precisely because of their comprehensiveness they were never implemented.

Sustainable housing development in Funafuti cannot be achieved under present centralised growth trends and would necessitate decentralisation of government activity. Experience elsewhere in the Pacific, notably in Kiribati (Chapter 3), shows that even population decentralisation is unlikely to reduce dwelling densities at the centre without strict controls over population movement and incentives to enable the relocation of economic activity in a way that has not been chosen in any Pacific country. Moreover, even attempts to decentralise urbanisation from the two main islets to elsewhere on Funafuti atoll have failed (Connell 1999). In such a small country the expertise and will to expedite changes, within or beyond the housing sector, have been largely absent.

Failures of housing strategy?

National housing policy has been established and under discussion for at least two decades in PNG and Fiji, but has only recently attracted government attention in Polynesia and Micronesia, with full-scale housing studies first being undertaken in Tuvalu and Samoa in the early 1990s (EDU International in association with GHK/MRM International 1992a, 1992b, 1992c; Clarke 1993a, 1993b) and Kiribati in 1996 (Kiribati Housing Corporation 1997).

That official interest is so recent points to relatively adequate housing conditions until the 1990s in comparison to earlier housing pressures in Melanesia. In Samoa the 1992 housing review became the catalyst for government to consider both the introduction of physical planning legislation and the possible establishment of a town council in Apia. The situation in Samoa illustrates the relative futility of sophisticated data analysis, which provides comprehensive information about the housing sector but has not led to specific strategies because of government unwillingness to consider the wider context. (Similarly in Tuvalu, a detailed consultancy report provided comprehensive national recommendations but in a national political context where these were almost impossible to implement.) The second case study outlined the way in which the Kiribati Housing Corporation identified a series of draft housing strategies, but Kiribati lacked the financial and human resources to implement them. Each of the three states has moved hesitantly towards the development of appropriate housing strategies, but without the effective priority, resources or will to achieve them, despite detailed analysis.

Housing strategies in Samoa

The Samoan housing study of 1992, an attempt to develop a national housing strategy, called for a set of 'recommendations to the Government as to its appropriate roles, objectives, policies, priorities, programs, targets and resources to facilitate the provision of housing and housing finance in the country, with special emphasis on provision of affordable housing to low income families on a sustainable basis' (EDU International in association with GHK/MRM International 1992a: Appendix 1-A). Hitherto the housing sector was scarcely touched by any form of government intervention. The small and undeveloped formal housing production system was involved in only one major project in 1992, and overall only 500 units per year were being constructed. Traditional homes (*fale*) were rapidly being replaced by European designs, which made up 71 per cent of the housing stock. Although more desirable in an urban setting because of new needs for privacy and security, this necessitated modern engineering and architectural design standards in housing construction. Most urban residential plots in Samoa are large, typically a quarter acre (1,000 sq m) and consequently housing densities are low. Development control has generally been non-existent in Apia, with planning permission not required for construction of *fales*, though permits were necessary for modern houses. Building inspections were rarely made and certificates of completion not demanded. Building permits show a tiny housing market with only 200 approvals a year in the period 1987–91. Most new dwellings were built by informal methods, with more than a third constructed by the occupants or by unpaid labour; a fifth were constructed by paid workers without a loan; and only 1.4 per cent were purchased ready-built. European-style houses were similarly constructed in a semi-informal manner, with various workers paid in cash or through reciprocity, quite different from housing production in most industrialised societies (EDU International in association with GHK/MRM International 1992b: 28).

Unusual increases in cyclone activity in the late 1980s and early 1990s inflicted extensive damage on Samoan housing stock to the point that individually organised self-help was unable to cope. When added to growth and increasing segmentation of the housing market in Apia, this created hitherto unprecedented pressures on residential accommodation. Further pressures came from a demand for modern rental accommodation by expatriates attracted to work in the public sector and from Samoans returning from employment overseas. Heightened demand raised the price of construction materials and building costs (also exacerbated by a high import content). Moreover, building skills were in short supply in a country which has traditionally exported the best of its trained workforce. Land supply was an important issue in flood-prone parts of the town. Illegal squatting did not exist but some customary areas and church lands experienced unacceptably high levels of overcrowding, and some road reserves were occasionally used for sleeping out. Such problems were reflected in rising freehold land prices and an increasing lack of control over urban development.

The production and management of housing in Samoa are affected by four main sources. The first of these is the Housing Corporation created in 1990 in response to market demands for modern forms of housing: an independent statutory body with the principal objective of lending mortgage money on security to those who would not gain such assistance through normal commercial banking loans. The Corporation had approved some 782 housing loans between 1990 and 1992 but waiting lists were long and lending activity had been curtailed because of a lack of funds. Second, the Public Works Department (PWD) has oversight of public infrastructure development and building control, regulates building construction and provides technical assistance to low-income families on building design and construction. A third body, the Department of Lands, Surveys and Environment (DLSE), has a long-established responsibility for land affairs and the environment. Finally, the private housing sector, small in both the number of households involved and financial size, is none the less the principal source of housing for middle- and upper-income groups. The components of the private housing sector are the banks, building industry professionals, developers, freehold landowners and suppliers of building materials. As in many other small, developing countries, the building construction industry is made up of an overseas-owned and -based modern sector which exists alongside just one significant local contractor. Few private housing projects exist but formal housing production is growing through the development for rent of some modern apartment buildings in Apia. However, in 1992 professional contractors could only provide 40–50 houses a year or only some 10 per cent of demand (EDU International in association with GHK/MRM International 1992c: 23), a situation that has not subsequently changed, though the government has sought to increase its reliance on a revitalised private sector for most new housing initiatives in Samoa. Housing has been given greater government priority, though urban development planning in Apia has yet to be established. Although the Samoan government has begun to consider housing strategies, policy formulation has effectively been left to a small housing authority with few financial and skilled resources to develop and implement appropriate housing projects.

A national shelter strategy in Kiribati

Around a third of the Kiribati population of 90,000 live in South Tarawa at an average density of over 1,800 persons per square kilometre. The town has experienced an annual population growth rate of about 3 per cent in recent years, a considerable proportion of which is made up of migrants from other islands who find accommodation with their extended families. Homelessness is non-existent but households are very large and there are increasing difficulties in providing clean water, sanitation and secure urban land title (Kiribati Housing Corporation 1997). Pressures on land in South Tarawa and the small national economy, which has a public sector providing almost 70 per cent of total employment, have made substantial investment in housing stock and support services almost impossible. Housing strategies must work within these and other constraints to have any chance of success.

Housing Corporation strategies were separated into a first phase of eight short-term measures for the period 1998–2000 covering the immediate needs and demands, and longer-term measures for the following eight years (ibid.). Although most strategies are difficult to implement, they extend well beyond the confines of the dwelling environment to embrace land reform, water and sanitation, finance, the identification of urban regeneration areas, and community participation, before more specific approaches to housing itself. With so many diverse themes, the Kiribati housing strategy is very dependent on continued external assistance in the form of infrastructure finance, professional skills and institutional strengthening; hence, although it provides a good example of housing philosophy and the evaluation of housing in a wider context, actual policy development is limited and it is very far from contributing to national sustainable development.

Managing housing stress

Many housing problems are ubiquitous. Housing agencies, where they exist, are incapable of embarking on more than a few token public housing schemes; even these are largely financed by international aid with minimal local initiative and responsibility. Most fall well behind target due to escalating costs or problems in acquiring urban land. Even site-and-service schemes may not be successful because of difficulties in land acquisition and the provision of infrastructure, and rarely meet the needs of the poorest in the urban population. Both equity and efficiency objectives, in the few instances they are identified, are more likely to be achieved by the upgrading of existing migrant settlements than by costly attempts to increase the supply of new houses. State housing organisations have failed to satisfactorily provide new housing stock for the relatively poor, primarily because they have rarely tried.

The identification of national housing policies cannot be separated from the problems of urban economies, where labour costs – and therefore housing costs – are very high. Squatter settlements are growing in size and significance; symbolically, settlements in Honiara have changed from being called 'temporary housing areas' to 'traditional housing areas', a measure both of their permanence and their few modern facilities. Though some of these settlements and others elsewhere in the region have existed for more than 40 years, they have received few services from governments and municipal authorities. If housing provision could be integrated with more effective urban planning and greater attention be devoted to the urban poor then real economic gains would ensue, especially if combined with government housing authorities becoming facilitators in the housing process rather than producers of dwellings.

Remarkably little is known about housing stocks and household needs in most towns. Ironically, only in Kiribati, the least prosperous country in the region, have adequate national attempts been made to understand and manage the housing process. External consultancy reports have been even more detailed, as in Samoa and Tuvalu, but have rarely met local aspirations and ability. In Micronesia governments have largely ignored housing policy and there has been no more than reluctant government involvement in regulating the housing sector in Polynesian countries (despite the multiplication of, and lack of coordination between, agencies, as in Samoa). Few planning controls are relevant to local housing design or housing projects and no requirements are in place to obtain planning permission prior to development. Where there are building codes they are not adequately policed. In Tonga, Tuvalu and most of Micronesia there has been no government organisation solely concerned with housing issues and little specific financial support for housing construction. No public housing exists in any Polynesian or Micronesian state other than for some public servants. Past housing advisory plans rarely took note of either indigenous culture, the realities of conservative and hierarchical political systems or the need to encourage sustainable forms of development. Whilst Samoa, Kiribati and Tuvalu have moved towards new housing strategy studies, the dilemma remains that to improve the housing situation substantially will require overall changes to government policy affecting national approaches to the question of urban development. Decentralisation, land reform, effective land use planning and good environmental practice are just some of the areas of public policy which could indirectly improve the urban housing situation, rather than any combination of more specific measures. Their absence is a measure of the challenges that remain.

The urban environment

Good environmental practice has been difficult to achieve in the urban areas of the Pacific, despite official data suggesting high levels of sanitation in some urban areas. In Vanuatu, for example, 90 per cent of the urban population are reported to have access to safe water and to adequate sanitary facilities, despite the leading causes of illness being infectious diseases and skin complaints caused by poor sanitary conditions and personal hygiene (Asian Development Bank 1997). However, such high rates of 'adequate' sanitation are considerable overestimates, with many people living in informal housing in and around Port Vila excluded from such statistics. The close connection between worsening urban physical environments and increasing urbanisation is widely evident; hence there are doubts about the capacity of governments in the Pacific to halt deteriorating living conditions or slow the pace of unmanageable urbanisation and, in the absence of adequate data, especially on informal settlements, even recognise that real problems exist.

Inadequate water supply, sewage and solid waste disposal problems are among the more obvious negative outcomes of urbanisation and are closely associated with the housing situation. Frequent complaints from householders about government inadequacies in dealing with them are found in newspaper letters and editorials, and the health status in several urban areas, where water-borne diseases are common, is indicative of the severity of these problems. Governments have provided public subsidies to induce private industry, yet make inadequate provision for the supply and management of essential urban infrastructure and services. In most urban areas at least 40 per cent of water supplies are lost through broken and leaking pipes and 'clandestine connections'; Honiara and Tarawa both lose at least 60 per cent of their water in this way, compared with global averages of around 20 per cent (Johnston 1999). Such problems are accentuated by the lack of effective land use planning and uncoordinated, fragmented urban development. In towns such as Apia residents lack both the protection of a legally enforceable urban plan and an elected local government representative capable of taking their concerns to a municipal council. Ironically, in Apia, even the South Pacific Regional Environmental Programme (SPREP) headquarters has been 'obscured by a nearby tar-making plant which has been spewing out clouds of soot for months. Residents living near the industrial zone have not had much luck in complaining to local health authorities or the environmental body as nobody has the authority to step in' (*Islands Business Pacific*, December 1993: 13). In such circumstances people were forced to target remote and not particularly accessible national government ministries for any response.

Such dependence on distant authority to provide most of the modern essentials of urban community life is usual in Pacific towns and follows the creation by colonial administrations of dependent urban outposts where indigenous residents were not encouraged to involve themselves in urban affairs. Yet the experience of cities in much of the developing world indicates that improvement in the quality of the urban environment depends increasingly on the efforts of townspeople, though distinctions should be drawn between broad city-wide environmental issues, like the supply of water and energy, that are mainly a governmental responsibility, and individual household activities like energy conservation and domestic waste recycling (Drakakis-Smith 1995). Both are vital to the health of the urban environment and form part of the 'Brown Agenda' (World Bank 1993a) of urban environmental issues: water supply, the treatment and disposal of solid and liquid wastes, air and water pollution, and traffic congestion, whose management is central to the quality of urban life. The case studies discussed here are representative of current conditions in the Pacific, though they are drawn primarily from the Polynesian towns of Apia and Nuku'alofa.

Urban water supply development

Progress in water policy implementation in the Pacific has been very limited, despite its obvious and growing importance. Indeed, when an international five-stage classification was developed that recognised the importance of several key components of progress towards water supply development – the existence of an internal champion to act as a catalyst, political will, an adopted approach or model as the basis for a national action plan, the presence of suitable institutional arrangements and, finally, capacity for implementation – only the first incipient stage (no champion to initiate water sector development and no political will) applied to Pacific island countries, alongside Myanmar and Cambodia (Arriens *et al.* 1996). Despite water being the most basic need, supplies are rarely adequate in either quantity or quality. In extreme cases like South Tarawa, where most of the population rely on well water for drinking and bathing, not a single well (out of a sample of eighteen) has met the WHO's standards (Christensen 1995: 41–2), whilst drought diminishes the quantity and reduces the quality of what is available, and inadequate or non-existent sanitation systems pollute the groundwater, as they do in most atoll environments. In both atoll states droughts in 1999 resulted in absolute water shortages and appeals for emergency seawater desalinators, as water consumption fell close to a 'lifeline' level of supply.

Urban water supplies are rarely adequate, and in several towns – particularly in Micronesia – supplies cannot be guaranteed. Problems have intensified as the public works systems installed in colonial times creak and rust into obsolescence. Leaks, storm overflows, shortages in dry weather, ageing and damaged assets (including tadpoles in the water pipes in Honiara), illegal connections, and failures to collect revenue have all exacerbated the basic problems. In Port Moresby, where rainfall is very low, water supplies are rarely adequate, and the duration, timing and quality of supply is uncertain; hence, in urban villages and squatter settlements especially, finding water has become analogous to hunting. Whilst some businesses have their own private water trucks and a new entrepreneur, the 'water merchant', has appeared (sometimes accused of sabotaging supply lines in pursuit of further 'privatisation'), consumption has nevertheless fallen and hygiene worsened. Neither the public nor the private sector has solved the problem, hence urban residents have sought self-reliance through the construction of local reservoirs (Gregory 1997), a drastic and unlikely solution in a drought-prone area.

The reasons for this neglect vary but a critical problem is the specific manner in which water has been provided:

> In the majority of countries . . . water has been treated as though it were available in unlimited quantities, and supplied at zero or low cost to consumers who resent the idea of water as an economic resource. Consumers, abetted by their governments, have clear notions of their water 'requirements', and the task of water authorities has, until recently, been seen as supplying those needs, with cost a secondary consideration.
>
> (Winpenny 1994: 2)

Pacific island countries have however targeted the supply of clean and affordable water in urban areas as a major priority in national development plans and sought foreign aid to rectify worsening conditions in most towns. But the dominant 'entitlement syndrome' (ibid.), which relies on the search for more supplies to meet rarely questioned or investigated demands, is unsustainable.

Water supply solutions alone will not work in the Pacific (or anywhere else) because hydrological limits to water reserves in most countries impose 'water stress' when annual renewable supplies fall below a reserve of about 2,000 cubic metres per person; the environmental costs of water supply schemes incurred by depletion of aquifers, damming rivers, destroying wetlands and disposing of wastewater are usually considerable; and various financial constraints characterise undercapitalised utility companies that cannot cope with the demands of anticipated urban growth. The main reason for the weakness of the performance of utility companies are failures in pricing and cost recovery that only manage to pick up one-third of the full economic cost of the water supplied, as high proportions of bills are not paid (ibid.; Hunt 1998). Urban water supply in the Pacific involves all these factors, several of which are evident in Apia and Nuku'alofa.

Planning Apia's water supply

Much of Apia receives water from a piped supply and its provision has long been the responsibility of the PWD. High leakage and wastage emphasise considerable difficulties in providing adequate supplies. Apia water consumers pay a small annual charge for their supply, hence draft legislation was introduced in 1992 for a user-pays system for metering all government offices and private homes, to rationalise use rather than increase revenue. Samoa, and especially Apia, had what is probably the highest water consumption rate in the Pacific – over 600 litres per capita per day – three or four times more than in most countries (and very substantially above the approximate 20 litres per day of Funafuti in Tuvalu). Most of this was used in Apia but the high level of consumption was primarily a result of considerable wastage, alongside the lack of water meters, some sabotage and illegal tapping of pipes. Vandalism, leakage and breakdowns also led to low water pressure and a reduction in water quality.

This placed extra burdens on the economy by channelling more funds into building adequate structures to meet these high usage rates, and lack of equity, as the national budget effectively subsidised inefficient users. Provision of urban water supplies has been both difficult and expensive because of the virtual absence of surface water, despite high rainfall.

The administration of water supply and sewage treatment in Samoa was transformed in the 1990s with the establishment of a new Water Authority and piped sewerage schemes in Apia, to improve environmental sanitation and minimise health hazards and the pollution of inshore marine waters and surface water. The corporate Western Samoa Water Authority, replacing the Water Division of the Department of Public Works, was an alternative to full privatisation of the kind introduced in other parts of the region such as Vanuatu. The PWD Water Division had been criticised for low productivity, poor billing, technical problems (which led to half of all water being unaccounted for and wasted) and a poor public image. Billing, for example, was carried out once a year by means of a register of water connections. In theory consumers who did not pay within six months were disconnected but, in practice and partly because of an unreliable register, less than half of all consumers actually paid their water bills at all. From some 15,000 house connections, only a few hundred were disconnected. At least two severe constraints confronted the PWD in its administration of water supplies and prevented it operating efficiently. For political reasons, the water rates approved by the government were completely irrelevant to any cost recovery objective, therefore the Water Division became totally dependent on government subsidies and good will; moreover, through the government's tight control on personnel recruitment and salaries, it was unable to develop and implement an appropriate human resources policy (Robert 1993: 16). Although the average monthly family bill successfully charged by the Electricity Power Corporation in Samoa was between WS$30 and WS$40, there was a negative reaction to, and a reluctance to pay, a more realistic and equivalent charge for water (ibid.: 42), since the perspective of most Samoans is that water is a natural resource which satisfies vital needs and should be freely available in abundance. Lavish use of water, including wastage, and the failure of the Water Division to analyse the population data needed to budget adequately for operation, repair and maintenance requirements (GKW Consultants 1989: 1/11) emphasised the PWD's difficulties.

For a government expecting the new water administration to 'break even as soon as all required rehabilitation programmes are completed' (Robert 1993: 44), a more appropriate tariff structure, enabling a significantly better return than the average flat rate paid by domestic consumers, was necessary. The chief issue, beyond extending meter connections to all urban and major rural consumers, was whether to implement either a 'hard' option of recovering operation, maintenance and depreciation costs or a 'soft' option of recovering only operation and maintenance costs. There was a fear that a 'two-speed' water service might eventuate which would consist of a free, intermittent and unsafe water service, where the Authority was not involved, and a costly supply where it was. Costs of associated sewage and waste water disposal facilities in central Apia would also be high, with the 1992 construction cost for a screening plant and sea outfall estimated at WS$42 (US$14) million (ibid.: 6). A decision on establishing a modern water supply and sewerage system was consequently deferred. Moreover, the Water Authority was obstructed in its efforts to impose a new user-pays system on the 45,000 residents it served; attempts to introduce meter use were initially met with widespread vandalism and tampering (Storey 1998b: 69). Consumption eventually dropped to 350 litres, still more than twice that in Vanuatu, when meters were eventually introduced in 1998 (Anon. 1999), but sustainable water provision remained out of reach.

Solutions to Apia's growing problem with urban services delivery and management conflict with traditional attitudes towards individual and collective behaviour which discourage limitations on personal freedom. An essential first step under these circumstances was to raise public awareness through better education and information, though the Authority was also advised, for the sake of good public relations, to avoid discussion of water quality and health hazards as long as raw water was supplied to consumers (Robert 1993), a measure of the manner in which water quality and delivery had become the subject of public scrutiny.

Water quality in Nuku'alofa

Considerable problems affect water quality in Tonga, if not its supply. Groundwater in Nuku'alofa is polluted from the outflow of septic tanks, which serve 70 per cent of households and most commercial establishments, and from pit latrines. Negative pressure in the reticulation system, combined with inadequate chlorination, has produced a potential problem of faecal contamination of urban drinking water, described by a foreign technical consultant as 'an environmental disaster waiting to happen'. About 85 per cent of the national population use groundwater, whilst the remaining 15 per cent rely on rainwater catchments; this is broadly the situation in Nuku'alofa where brackish groundwater is used for washing, cleaning and flushing toilets in those parts where piped supplies are available and rainwater is used for drinking.

Water supply has been the subject of major investigation in Tonga involving the preparation of a series of reports and master plans (PPK Consultants in association with Riedel and Byrne 1991, 1992a, 1992b, 1992c; Pulea 1992). Concern had developed over damage to freshwater lenses by overpumping, resulting in sea water intrusion, poor enforcement of existing water regulations, and deficient legislation which allowed only minimal control of water pollution and did not set adequate standards (Tonga Central Planning Department 1991: 248–9). The key problems are different from those in Samoa, owing to the very different geological circumstances of Tongatapu, a raised limestone island without surface water.

Although the four urban water systems of Nuku'alofa, Neiafu, 'Eua and Pangai/Ha'apai are controlled by the Tonga Water Board, two other authorities have important responsibilities. The Ministry of Health is responsible for rural supplies and the Hydrogeology Section of the Ministry of Lands, Survey and Natural Resources has responsibility for pumping, water testing and managing groundwater resources through its drilling programme (Pulea 1992). However, Tonga has a Water Supply Master Plan which, in Nuku'alofa, has made several recommendations for new programmes, including rigorous leakage detection and control; public education for water conservation; and major institutional strengthening. The existence of a Water Board assisted in making Tonga's water delivery and cost recovery system more efficient than that of Samoa. Almost all consumers are metered, meters are read on a monthly basis and rate notices are hand-delivered to consumers. Charges are increased in progressive steps according to consumption and although the number of water disconnections has increased annually, the figure for 1989 was only 96 out of 4,840 domestic meters in the urban system. Average consumption for the town was 831 litres per connection a day in 1989, around half that of Apia, with about a quarter of supplies disappearing in unaccounted leakages.

The total population of Nuku'alofa is forecast at about 43,000 by 2011 and a key difficulty for water and other services provision is the likelihood that residential development may continue in an unplanned way. In the words of the Nuku'alofa Water Supply Master Plan: 'The current demand for land is being satisfied through opportunistic development of garden allotments (*'api uta*) in and around the Nuku'alofa area and by Government subdivision of Government Estates in the low-lying areas of Popua and Sopu' (PPK Consultants in association with Riedel and Byrne 1992b: 4). Moreover, only two potential sources of land for new housing development exist in Nuku'alofa through the 'redevelopment of existing but vacant leased land in the urban area; and production of new land by reclamation in appropriate areas' (Kinhill Riedel and Byrne in association with PPK Consultants 1993: 6). Both strategies to increase the supply of residential land will have significant environmental effects on the Fanga'uta Lagoon, and either raise the cost of water provision or reduce its quality in outlying areas.

Water solutions?

Securing sustainable water supplies is an international goal which has received considerable attention in the Asian region, but little in the island Pacific. A national focus for water management is necessary, whilst delegation of water services to specific agencies requires an increase in the autonomy and accountability of service providers. Achieving sustainable water use means developing incentives, regulation and awareness, and requires better water information, consultation and partnerships, alongside capacity building, monitoring, evaluation, research and learning (Arriens *et al.* 1996). Such principles, evidently true of housing, might also be the basis for the provision of most other services – from electricity to transport – within the region.

Whilst adherence to such idealistic principles would result in major improvements to water management, small governments in the Pacific have had great difficulty in implementing the most basic aspects of public policy. One response to the shortage of technical skills and managerial experience has been to transfer responsibility and ownership of public utilities to private or corporate control. As elsewhere, privatisation and corporatisation have been challenged as leading to reduced equity and fears that the poor will be disadvantaged:

> It is safe to predict that turning water into a commodity will prove controversial and will draw deeply on a government's goodwill and credibility with its citizens. Governments should, however, seek every opportunity to remind their citizens of the alternative, which in many cases will be a grim scenario of growing and eventually disastrous water stress.
>
> (Winpenny 1994: 110)

Moreover, decreasing water consumption by imposing costs reduces income to water authorities and provides extra incentive for illegal connections (Hunt 1998: 135), already rife in the region.

Development conditions in Pacific towns and cities vary and government responses to the provision of urban services reflect these differences. In Suva, for example, progress has been made in creating and maintaining a reliable water supply. Since 1993, the water and sewage management responsibilities of the Fiji Ministry of Infrastructure and Public Works have been transferred to a semi-privatised corporation. The water supply situation in metropolitan Suva is good for those in formal accommodation with over 95 per cent of formal dwelling units in the city having water connections (Olsen 1991) but 40 per cent of the urban population lack access to water, proper sanitation facilities and waste collection services (World Bank 2000a: 8). Provision of water in Suva reflects the favourable water resource situation in eastern Viti Levu and work on the supply system since the city's early establishment. Such reasons do not fully explain why other Pacific capitals with similar attributes, such as Apia and Honiara, are not nearly as well served. The condition of urban utilities clearly depends on a variety of socio-economic, cultural and physical factors, including the marked differences in inherited colonial institutions evident in the contrasts between British, French (in Vanuatu) and American (parts of Micronesia) models of local government. In Port Vila a distinctive French colonial influence has led to a more privatised provision strategy. Before independence in Vanuatu the French company Société Lyonnaise des Eaux had supplied some water via an *affermage* system whereby 'water supply companies have a local monopoly, but the tariff is determined by a public authority' (Roth 1987: 252). Under the *affermage* system now operative in Vanuatu the public authority constructs the system and contracts out operation and maintenance work, collection of charges and consumer relations to a single private operator.

Regardless of private or public provision of infrastructure, several key reforms to public utilities must be met in Pacific countries for improved service provision: service providers need clear and coherent goals focused on delivering services; their management should be autonomous; they require financial independence; and there must be greater participation by users (World Bank 1995a: iii; Arriens *et al.* 1996). In the only comparative study of urban environmental sanitation services in the Pacific, the World Bank (1995a) recommended that an autonomous agency for water supply was the most practical option for improving institutional capacity in the short term.

Managing sewage, garbage and solid waste

Throughout the Pacific examples of severe environmental stress indicate that unsustainable living conditions are commonplace. Extreme problems are found in the atoll states of Kiribati, Marshall Islands and Tuvalu, where high-density settlement on limited land areas has increased the need for urgent solutions to the disposal and treatment of urban wastes. Sewage is the most noxious, but usually least visible, of these. Few towns possess proper sewage treatment facilities. In Melanesia, for example, only Honiara and Suva possess a reticulated sewage collection and treatment system. The small township of Maprik in PNG had to be 'closed' for several weeks in 1992 when its whole system overflowed. What was adequate for the small towns of colonial times has become inadequate for the larger towns of a new era, to the extent that SPREP stated that 'disposal of solid and liquid wastes (particularly of human excreta and household garbage in urban areas), which have long plagued the Pacific, now emerge as perhaps the foremost regional environmental problem of the decade' (quoted in Hildebrand 1999: 24). Progress has been made in some towns: South Tarawa has had a reticulated sewerage system since 1993 (though most urban residents do not use it, but use the beach instead), and on Funafuti septic tanks are mandatory for every residence (Hunt 1998). But only about a quarter of the population in Suva, usually the most advanced city in the Pacific in terms of infrastructure, has a sewer connection. Treatment plants, where they exist, operate at full capacity and are rarely able to prevent some pollution of marine environments. Sanitation has been neglected in the Pacific – usually out of sight (if not of other senses) and out of mind – or is unsuccessful because of inappropriate technology, lack of involvement of the supposed beneficiaries and lack of clarity within government over who should provide, and pay for, the service. Neglect may be costly; the 1999 cholera outbreak in Pohnpei cost ten lives, closure of schools, high costs for eradication and hospitalisation, loss of income from tourism and the threat of a Japanese ban on tuna imports (Howorth 2000). Sanitation in squatter settlements, and overall equity of provision, is usually problematic. In Blacksands, there were no flush toilets and only pit latrines, which smelled bad and usually filled up within a year, hence new spaces for pits were becoming difficult to find. Similarly as many as forty people were using the same well, which was sometimes polluted as toilets were close to wells and accessible to flies and mosquitoes. A few households 'had rainwater tanks distributed by politicians at election time' (Mecartney 2001: 179). More

Plate 7.4 Matavu settlement, Suva, Fiji. © John Connell

generally landowners were uninterested in turning settlements into serviced suburbs where tenure appeared more secure.

The scale and cost of solutions are considerable. The cost of overcoming the backlog of unconnected households and enterprises in metropolitan Suva alone would be more than F$60 million (US$38 million), if dealt with through a conventional sewerage system, a vast sum for even a relatively developed island state. A cheaper and more balanced approach to remedy existing deficits was therefore recommended which combined conventional reticulation with additional participation by local consumers in sewer maintenance, potentially resulting in considerable cost savings (World Bank 1995a). In Port Vila the importance of the tourist industry to the urban economy, as well as local health considerations, have heightened concerns about pollution of the lagoon and harbour. An effective response has been precluded because of divided responsibility for environmental matters between several authorities, resulting in failure to provide for adequate sewage disposal in the town. Weak urban planning was mainly responsible, though there were other important contributing factors such as a lack of finance for capital works and failure to resolve disputes over the release of more land for residential development. In every urban area in the Pacific sewage disposal has become a critical environmental problem.

Domestic solid waste generation constitutes a rapidly growing problem in the urban Pacific, since the volume of garbage is of the order of 0.7 kg per capita per day, but organic, biodegradable material still dominates its composition, followed by plastics, glass, metals, paper and cardboard. Volumes are steadily increasing and, in some urban contexts, the volume of 'modern' waste is perceived as a status symbol. Disposable nappies (diapers) are particularly numerous in parts of Micronesia. Medical, chemical and oil wastes and batteries present specific problems, demanding recycling. In small towns and islands recycling is at best economically marginal, almost every scheme appears to have failed (Morrison and Munro 1999: 238), and plastic bottles and cans litter urban landscapes – inadvertent advertising for multinational companies – despite attempts to impose additional costs on importers and distributors. However, where volumes are substantial, such as bottles in Port Moresby, a highly organised scavenging industry has grown up, and as many as 90 per cent of bottles are recycled (Hunt 1998: 146–7). Only aluminium cans are similarly recycled. Entrenched disposal habits, inadequate disposal points and irregular collection otherwise enhance the harmful environmental impact of the rapid rise in non-degradable garbage.

Garbage collection systems have been unsuccessful in providing efficient, effective and economic removal of wastes from households, offices and commercial premises, partly through lack of public willingness to pay for a service that is often inadequate, the lack of effective pricing structures, and inappropriate equipment, such as sophisticated compactor trucks with heavy maintenance demands. Only in South Tarawa is there a realistic collection fee structure and the sole garbage collection service in the region that does not run at a loss (Morrison and Munro 1999: 238–40). Landfill sites are inadequate in size, often poorly located (in mangroves or on lagoon foreshores) and unsupervised; scavenging is common – especially in Port Moresby, where some dumps, such as Baruni, now effectively support small populations. Leaching from dumps close to the sea is not uncommon, such as the large Lami dump that has contributed to heavy metal pollution in Suva harbour, one of the most polluted in the region. Incineration is a common informal means of garbage disposal in Melanesia, so much so that urban Melanesians have sometimes been regarded as pyromaniacs; for Trobriand island migrants in Port Moresby 'Smoke is the *sine qua non* of gardening success. . . . However, in the city, where the "weak" soil needs it most, smoking causes problems' (Battaglia 1986: 41) as smoke hazes contribute to respiratory infections. Squatter settlements rarely benefit from garbage collection hence the negative environmental impact of failures of waste disposal alongside other failures of infrastructure provision are particularly severe. The environmental situation in Apia and Nuku'alofa can be examined in more detail, as representative of the urban environmental management issues of Pacific towns.

The urban living environment in Apia

Most of Apia's sewage is currently disposed of in backyard pits or septic tanks, leading to increasing health risks. Effluent from small industries is discharged directly into lagoons, as is much of Apia's sewage. Open drains carry sullage and a variety of toxic wastes from such services as dry cleaners, petrol outlets, the fish market and the national hospital into the main harbour, which is a recreation area for many children. Major pollution problems in Vaiusu Bay to the west have included solid waste disposal, bacterial contamination and the destruction of living coral and mangrove areas (Klinckhamers 1992). The biodiversity and natural resilience of Apia's coastal systems has been significantly reduced through human actions, including over-fishing, reef blasting and dredging, with the result that the coastline remains vulnerable to erosion, despite the construction of a new sea wall. Existing environmental regulations are generally not enforced because cooperation between government departments and village organisations is rarely possible.

Several problems of urban environmental management are both visible and difficult to resolve, such as dealing with the swampy areas at Saleufi, Taufusi and Fugalei. Chronic flooding in Apia has followed construction of the seawall, suggesting inadequate environmental impact assessment in the planning process. The local press has described the situation graphically:

> It began when they reclaimed Apia's waterfront raising the sea level so that waters coming down the drains could not escape quickly into the ocean. This was aggravated when the hills were recklessly cleared for farming . . . so that much erosion took place when it rained. It still does. But as a result of all these floodings . . . the low-lying areas in the middle of town are swamped every time it rains – now Apia is once again the 'ugliest little town in the world' when flooding buried the centre of the town under muck and slime yesterday morning.
>
> (*Samoa Observer*, 7 February 1992)

Factors creating problems for urban environmental management include poor drainage and periodic cyclone damage; large areas of customary land; weak and disorganised institutional management and lack of legislative powers; and a small fiscal base from which to initiate infrastructure improvements. Pit latrines are not unusual in Apia precisely because there is inadequate finance to develop more complex schemes. Most of the urban population has a piped water supply but this is not always adequately treated and persistent low pressure disrupts supplies. Water supply problems, rubbish and waste disposal are prominent causes of complaint among Apia residents. In 1992 high coliform counts were found at the Afiamalu water intake supplying parts of Apia, with pollution from animal or human sources in the catchment, where there was no legal means of enforcing protection (ibid., 16 April 1992). Inadequate formal controls exist to cope with such problems. Environmental legislation, a relatively new concept in the Pacific, here as elsewhere has been slow to be enacted and, more importantly, implemented.

Installing a reticulated sewerage system in Apia has become essential and although detailed plans were prepared in 1992, the unit cost of the proposed system was then US$1,430 per connection (1992 prices), considerably higher than in many other developing countries. Alternative low-cost options such as ventilated pit latrines for the peri-urban areas of Apia and a condominial sewerage system (involving a shorter grid of individually maintained smaller and shallower feeder sewers running through backyards), of a type pioneered in Brazil, would be suitable (World Bank 1995a) but costs have deterred even such appropriate schemes.

Dealing with waste in Apia and Nuku'alofa

Solid waste disposal has posed health and environmental problems throughout the Pacific. Privatisation of garbage disposal, for example, has failed in Apia and service standards are erratic and unsatisfactory (Connell and Lea 1995: 120–1). Basic good management procedures, even when established, are difficult to keep in place, hence expectations about the privatisation of urban environmental infrastructure are misplaced. In this case, at least, management was equally inefficient in both public and private sectors.

The privatisation experience in Latin American and Asian developing countries suggests that while contracting out solid waste collection services offers the best opportunity to improve efficiency, governments usually have inadequate regulations to prevent the abuse of power by privatised monopolies. In Asian cities private sector operators have provided some services more efficiently but, to achieve this, responsibility needs 'to be shared and actions to be taken by many agents, including the national governments, local authorities, NGOs, communities and the private sector' (Lee 1997: 159–60). Decisions about significant community-wide change and improvement in Pacific towns can take a very long time to resolve. Part of the difficulty is the proliferation of public agencies, problems of defining responsibilities in what are sometimes new areas, and the predictable problem of coming to a decision where customary land may be involved. Seemingly straightforward decisions may be interminable in the face of a variety of entrenched interests and limited resources.

Municipal dumps in Tonga are also typically located in mangroves, where there has been dumping of paint residues, pesticide containers, batteries, plastics and general urban wastes in an unregulated manner. The Nuku'alofa dump at Popua frequently burns, causing pollution over nearby residential areas, and pesticides also contaminate the area (Pulea 1992). Urban solid waste disposal is managed by the Ministry of Health under special legislation but there is nothing to regulate what may or may not be dumped and no environmental impact assessments are required. Upgrading of the dump to the extent that it became a regional success story (Saw and Belz 1997) was shortlived, as adequate management was not sustained.

The most substantial and widespread problem of waste management in Nuku'alofa, and to a lesser extent in the smaller Tongan towns, is the disposal of sewage from latrines and septic tanks. Waste from public toilets, hotels, public buildings, industrial sites and private households is not treated. More than two-thirds of households in the capital use septic tank soil absorption for disposal, 20 per cent use pour-flush latrines and 10 per cent use pit latrines. Sludge is collected by Department of Works pump trucks and placed in small dyke enclosures to the east of the town (Pulea 1992). Some septic tanks have allowed surface leakage, especially in areas of the town below sea level, including the old Sopu lagoon area and Popua. Piped sewage collection was not recommended for Nuku'alofa because of its flat terrain, high groundwater table, low population density, low water use, vulnerability to disaster, lack of skilled personnel and high costs (Belz 1984: 32). A full sewerage system would be developed, it was thought, when population density and consumption increased substantially and in the meantime smaller and incremental improvements would suffice. Since then there has been little evidence of any investigation of even small improvements of this kind. Sewage disposal has never been perceived as a pressing problem in Pacific towns.

A future for housing and urban environmental management

Several existing policies require transformation in Pacific towns if substantial changes are to be secured. Housing development, for example, is crucial and central to provision of other services. Few other aspects of urban affairs so obviously reflect the demands placed on governments by urban growth and the effects of a changing economy. Sensible housing policies cannot be formulated without effective land management, well integrated environmental controls, good transport planning and appropriate infrastructure provision of all kinds, not to mention good understanding of building industry economics and national approaches towards finance and taxation. However, governments have tended to disengage from, as much as they have engaged with, housing issues. Consequently housing standards increasingly diverge between well-serviced detached housing (and apartments) for elite residents and a growing number of migrant settlements, especially in the largest Melanesian towns. The chief reason for the earlier existence of adequate dwelling conditions in many parts of the region was the late onset and slow pace of urbanisation, but that is no longer so.

Top-down and authoritarian ways of providing formal urban housing and infrastructure have been the norm but have had little success, with standards usually being based on those prevailing in industrialised countries, and authorities responsible for providing new infrastructure and services having 'little or no knowledge or interest in communities residing in the settlement' concerned (Cotton and Franceys 1994: 16). The dilemma is how to provide adequate low-cost housing and related services and extend them to growing urban populations, despite very limited financial resources and technical skills and where community support and involvement is minimal.

The extent to which community involvement might achieve more cost-effective and equitable provision of housing and services is debatable, where needs are complex, existing facilitating agencies and local governments generally inadequate, education standards are not high and hierarchical social structures exist. Although towns like Apia and Nuku'alofa are largely collections of urban villages possessing the characteristics of rural communities in their social and demographic structure, and are possibly appropriate places for greater public participation in urban affairs, they can equally be arenas for new tensions and conflicts over access to scarce resources. Indeed, the village structure may inhibit some forms of improvement best provided at a broader scale. None the less Pacific townspeople need to secure greater ownership of and involvement in urban affairs for progress to occur. Communities must be involved in all stages of improvement projects (where they exist); facilitating agencies should be supporters of the community, not owners or managers of programmes; while government agencies need to fulfil their tasks of motivation, facilitation, training and technical assistance (Briscoe and de Ferranti 1988). All of these, and particularly the last, are so far from the contemporary Pacific scene that they remain mere wishful thinking.

By contrast scarce public resources are used on unnecessarily expensive construction projects, where operating efficiency is low, and subsidies are needed for water and sanitation, leaving few resources to extend services to the poor (Briscoe 1992). Numerous studies have indicated that poor people want and are willing to pay for services of relatively high quality and reliability. It is appropriate to shift from a supply-driven orientation to a situation where households are offered choices. This will be difficult, since suppliers need incentives to reduce costs and be accountable to users, and is likely to require some degree of privatisation. Scarce public funds should be used for those services, such as treatment and disposal of wastes, that provide substantial communal benefits. There has been considerable international activity in the field of improving urban environmental services and one conclusion is that urban environmental strategies should:

> have an explicit focus on the problems of the poor; city-specific strategies should be guided by the configuration of key economic variables; solutions that are not heavily dependent on institutional performance may be necessary in the short run because of the organizational problems of complexity and synchronization; enhanced public awareness, consultation, and participation can improve environmental management; and [strategies should] pay careful attention to the selection of problem areas, their scale, and institutional capacity when designing interventions.
>
> (World Bank 1993b: ix)

But such appropriate suggestions are still quite remote from Pacific practice, despite the growing role of NGOs and churches at the community level in situations where urban local government is weak or non-existent.

In spite of considerable support and encouragement from international and regional development agencies and foreign aid donors, other than in Fiji, NGOs established in the Pacific are yet to have a marked effect on housing provision or the urban environment. In Micronesia, for example, local circumstances have seen the creation of many government departments, agencies and corporations with responsibilities for essential infrastructure services but there are few community-based organisations (CBOs) of a non-governmental kind that participate in the planning and delivery of urban services. The considerable potential of NGOs is exemplified in the work of the South Pacific Committee for Human Ecology and Environment (SPACHEE) in Fiji (Bryant 1993) and the O Le Siosiomaga Society in Western Samoa (Connell and Lea 1995), yet they have limited sway with government and little community participation.

A degraded urban environment characterised by the 'brown' effects of domestic and industrial pollution is not readily improved by initiatives focusing on one or more policy areas alone because of the need for integrated management of urban growth. This was realised at the 1992 United Nations Earth Summit and led to the preparation of environmental profiles and strategies for each Pacific nation, many prepared under the auspices of the South Pacific Regional Environmental Programme (SPREP), yet few have gone beyond the legislation phase. Towns occupy an important place in these studies but in scarcely any instance has it proved possible to link advisory environmental strategies with statutory land use plans, even where these exist. In the absence of such coordination, urban housing and environmental planning strategies occur through the medium of commissioned reports to aid donors, usually on a particular component, few of which are acted on, despite often being commissioned in response to crisis. Even under such crisis management the most politically sensitive and difficult housing and environmental service delivery issues are studied and reported upon in cyclical fashion by one aid agency after another, whilst the need for an overall planning and management structure remains as crucial as ever.

Part 3

Sustaining Pacific urban society?

8 Living in town

A sustainable future?

> Most of the literature on sustainable development does not mention cities . . . this reluctance to discuss sustainable development and cities is probably because many of those who write on environmental issues have long regarded cities with disdain, even if they live in cities. Among those who write about sustainable development, many probably consider cities as a key part of 'the problem'.
>
> (Mitlin and Satterthwaite 1996: 30)

Towns and cities in the Pacific are merely one part of the problem of achieving sustainable development in the region. Little now remains of the early colonial towns and the divisive policies associated with them, though their legacy is apparent in the enormous variation in the quality of the urban housing stock, the form of residential subdivisions, the low-density layout of most towns and cities and the policy contempt for the migrant poor. Over time, towns have grown more rapidly and a planned form has become less evident as informal settlement has become more significant, and expatriate populations have become minorities. They have quickly evolved into indigenous centres by offering residence and employment opportunities of many kinds to an increasing proportion of the regional population, and they have a growing dominance over national economies, as these have become more 'modern'.

Growth has brought social, economic and environmental problems as the cities have become more dynamic, complex and even elusive entities – as commuting takes hold – though their relatively small size and island locations have given them certain common characteristics. There are enormous variations between the Melanesian capitals and those of the atoll states, differences accentuated by distance, the lack of similarity of colonial or contemporary experience and the absence of any regional forums that might engender common planning and management approaches. Yet Pacific towns and cities are part of a world where local development issues have considerable similarities with those in other developing regions.

There are several common themes in Pacific urbanisation. The great majority of towns, and all the capital cities, are located in ecologically fragile coastal settings and some, particularly in the atoll states, are experiencing severe problems at present rates of population growth. The discharge of untreated sewage and other wastes into coastal waters, and especially lagoons where mangroves have often been cut back, has not only put fragile ecosystems at risk and increased erosion, but reduced the productivity of subsistence fisheries. Air pollution (from vehicles and burning waste) is a new phenomenon, as is noise pollution. Several towns occupy low-lying sites, parts of which are periodically swampy, posing environmental and public health problems. As urban and national economies have generally become more bureaucratic, despite 'downsizing', and partly fuelled by remittances, aid and the profitability of large corporations located there, the link to local and national economies has been partially broken. Where this has occurred on particularly limited and fragile ecosystems the environmental impact of growing urban population densities, without urban or national planning, has been severe. In most parts of Oceania the combination of a tropical marine location fringed by mangroves and sheltered (and blocked) by coral reefs and the lack of effective waste treatment and disposal capacity has led to pollution of harbours and waterways and damage to marine environments. Environmental damage – especially where it is linked to high urban population densities – is of obvious ecological significance but also of economic importance where it alienates or destroys land and sea areas of value for subsistence purposes.

Many towns were initiated in the early days of colonial trade and settlement in places unsuited to any but the most limited forms of settlement. Port Moresby owes its location to ease of access to Australia, rather than its links into the hinterland or its accessibility to agricultural land. Whilst some poorly sited towns such as Levuka and Salamaua were eventually bypassed, others remain on inappropriate sites where little land is available for development, thus creating high costs for urban development.

Customary land issues and social and legal problems of access to land are ubiquitous in Pacific towns. Island governments have largely been unable to deal with land issues. The ability of governments to manage urban growth rests ultimately on their success in controlling land markets, yet the majority of undeveloped land within urban areas otherwise suitable for future urban development is beyond the reach of regulation and management. This may in some contexts have accidentally contributed to sustainable development, where difficulties in urban land supply have acted as a constraint to urban growth (thus helping to lessen other, negative, impacts) and contributed to greater self-sufficiency on the part of urban residents. However, the existence of unregulated land in and adjoining urban areas has led to widespread informal settlement, even on the fringes of mining towns, compromised the ability of urban authorities to safeguard environmental assets, and raised the cost of infrastructure provision across fragmented towns. (Squatting in water catchments around Port Vila and South Tarawa, for example, has necessitated the search for more distant and more expensive sources of supply or costly relocation strategies.)

Although the colonial and political circumstances of Pacific island countries vary widely, and no single model of governance is likely to meet the needs of countries with great differences in geography and demography, none has achieved effective systems of government and management at the urban level. Some success has been achieved with conventional British urban local government in parts of Melanesia, such as Port Moresby and Suva, where there has been national government support and higher than average rateable income. Elsewhere the failure or absence of local government, even in relatively large cities like Apia and Nuku'alofa, emphasise political failure and uncertainties. As far as sustainability is concerned, the shift from 'government to governance' requires more involvement in urban management by individuals and community organisations. In Samoa NGOs such as the O Le Siosiomaga Society have for some time drawn attention to urban mismanagement at public meetings or in occasional publications (Vanderburg 1992), yet the role of community-based organisations (CBOs) in urban improvement has thus far been trivial, even – or perhaps because of – the questionable democratic record of (and the emergence of corruption in) authoritarian national governments in many parts of the region. In small societies numbering only tens or hundreds of thousands the influence of national government on urban affairs is enormous, hence empowering local authorities, or even responding to NGOs, implies changes to the basic organisation of contemporary society.

Urbanisation has occurred in a context of limited economic development where economic growth has been slight, and urban unemployment rates have grown, especially in Melanesia (where the safety valve of emigration is lacking). Rising unemployment has contributed to social problems, including crime, whilst low economic growth has reduced available revenue to national governments and urban authorities, weakening their ability to manage urban development. (In turn, failures of management contribute to slowing economic growth – through increased costs of, or inaccessibility to, infrastructure – and worsening social life.) Skilled human resources are scarce in most states. Resources for national development are thus limited and for urban development are particularly sparse despite urban biases in policy formation.

In this context of poorly planned coastal cities whose sites constrain certain forms of development, and where limited resources (of capital, land and skills) constrain other elements of management, urban societies have developed in an unusually fragmented manner. It is those societies, their social and economic structures and political organisation, that hold the key to contemporary urban development and planning. The provision of urban services has usually been piecemeal, ad hoc, poorly integrated or simply inadequate. But the colonial past, the indigenous past and the structure of contemporary urban life, have all contributed to this.

Cities of parts

Pacific towns and cities have evolved under difficult circumstances from colonial creations from which indigenous people were excluded or where their rights were severely curtailed to Pacific places. None the less, urbanisation was a belated product of the colonial encounter and the legacy of colonialism remains evident in the urban Pacific, more in terms of legislation (or the lack of it) than in the visual impact of suburbs and edifices. Many urban residents are still closely tied to rural, sometimes inaccessible origins; this has contributed to a lack of cohesion, and disarray and division are distinct features of the contemporary urban scene.

In the colonial era distinctiveness was regulated; assimilation, sexual relations and marriage between different groups discouraged; and multiculturalism wholly implausible. Thus, though civilised Europeans covered their bodies and Papuans were forced to wear loincloths in towns, notably Port Moresby, covering the upper part of the body was expressly forbidden for Melanesian men and women. Indigenous peoples remained, and were expected to remain, exotic. The government anthropologist in Papua, F.E. Williams, wrote in 1932: 'You can never be quite the same as the white man; and you will only look silly if you try to be. When we see a native in European clothes we usually laugh at him' (quoted in Inglis 1974: 6). Gambling and alcohol consumption were also forbidden.

Attitudes to houses paralleled those to clothes; when Motu and Koita villagers on the fringes of Port Moresby built houses with iron roofs, Williams wrote: 'We like to see the Motu and Koita houses. If you build one like a European copra shed it will not look very pretty' (ibid.: 7). Hence 'Both economically and in a larger sense the colonial order depended upon at once creating and excluding its own opposite' (Mitchell 1988: 164). Though divisions were less acute elsewhere in Melanesia and in Micronesia and Polynesia, urban separation – social, economic, physical and political – characterised the colonial order and lived on into post-colonial years.

Towns and cities, though now indigenous in ethnic structure, retain their colonial layouts and housing estates and have acquired airports, TV stations and multi-lane highways rather than pavements or bicycle tracks. The informal sector is inconspicuous, a victim of retained colonial legislation, lack of skills, small markets and sometimes superior earnings in the rural sector, even where it employs many urban residents. The hustle and bustle of Asian and African cities is absent, especially on Sundays, as Christianity and trading legislation combine to impose peace. There are few urban festivals, political rallies or strikes where crowds gather. Urban life is more of a personal phenomenon in Pacific towns. Although there are third or fourth generation urban residents, impermanence is still inherent in some aspects of urban life. Built into the contracts of many workers, including the prestigious occupations of mineworkers and bureaucrats, are annual return fares to village 'homes'. In such a context even long-established urban residents may have a rural home and a diminished stake in urban life.

Many urban residents, perhaps most in some Melanesian towns, were born outside the towns. The majority of urban residents speak their 'home' language, some Pidgin English (in Melanesia) and English (or French in Vanuatu) but most are more confident in their vernacular which possesses a cultural resonance and complexity that imported and recently constructed languages lack. Yet as duration of urban residence increases migrants are more likely to marry in town and, in Melanesia particularly, across language divisions. In Blacksands (Port Vila), as many as 45 per cent of all households were in inter-island partnerships and for them town was a 'neutral place' of residence, even if few regarded Vila as their 'place'. Their children, born in town, are the first to lose their supposed island affiliations; as one urban resident, a pastor from the Banks Islands married to a woman from Ambae, said:

> We mainly speak Bislama at home or English if we are helping the children with their homework. I try to speak to them in island language so they will understand it and my wife speaks to them in her language, so they can speak to her relatives when they visit. The children just speak to us in Bislama though, as it is hard for them to try and learn three or four languages at the same time.
>
> (Mecartney 2001: 203)

Over time creolisation increasingly becomes the norm for urban residents (Jourdan 1994). Most live with fellow migrants or their descendants, identify with their social concerns, and experience city life via a world of kinship contacts. Where employment is dominantly a male phenomenon, particularly in Melanesia – enabling males to cut across social ties in their workplace – and migration of women is mainly passive, social life in the city will remain something of a gendered world. To a substantial extent in the largest Melanesian cities, such as Port Moresby, men continue to carve out 'domains away from women and children – on street corners, in sporting teams, in drinking and pool clubs, in exclusive "men's houses" . . . and in *raskol* gangs' (M. Ward 2000: 234). Whilst friends were more important to men and neighbours to women, kinship continues to dominate urban social life, but especially for women (Young 2000), though their urban roles – notably as in the Gorokan settlement of Okiufa, where they act as 'tamers of men and guardians and investors of their husbands' income' (Zimmer-Tamakoshi 1998: 202–3) – may be anything but passive.

In many spheres access to employment is gained through kinship or *wantok* associations. Even elements of the 'modern' sector, such as the hundreds of buses that ply the roads of Port Moresby, are owned and operated by kin groups mainly from the Highlands. Few organisations cut across such social phenomena. Unions are few, rarely militant and primarily associated with the upper echelons of employment (such as bureaucrats and mineworkers). Class consciousness remains far weaker than ethnic affiliation and traditional hierarchical social structures are powerful. Other organisations – churches, sports clubs and so on – subsume social groups rather than cut across them. Few townsfolk are active or prominent members of church or sports groups; in Melanesia most such organisations have an ethnic base, even amongst tiny migrant groups (Strathern 1975: 164, 364 ff.; Battaglia 1986: 9). However, despite the links between churches and particular regions, they were one of the few means through which

Plate 8.1 Suva city centre, Fiji. © John Connell

urban residents from different areas came together at all. In PNG the urban gang is 'one of the few structures in which tribal lines are blurred in favour of larger social groupings' (Harris 1988: 47; Nibbrig 1992). Urban residents can live much of their lives as members of their own cultural group and most choose to do so. Ties with 'home' and kin are vastly more important than links to neighbours from other cultural groups, though these may be mobilised for particular reasons (Strathern 1975: 256–73). Even for those in formal sector employment, urban life is not always intended to be permanent; for most migrants urban residence is a 'rural-orientated' strategy, designed to generate income and prestige for a successful return to the security of rural life where land rights exist. The satisfactions of urban life are hard to find and its rewards may be few, as expressed by a Trobriander in Port Moresby:

> Working in the office, making money, pushing papers. . . . I sign other people's work. You don't get the kind of wisdom you require, personally. The personal satisfaction. To see the results of my personal achievement . . . the feelings in it. . . . This feeling of Trobriand pride has been entrenched in us as little kids. Thinking, why can't we do this here in Moresby? It belongs to Kiriwina.
>
> (Quoted by Battaglia 1986: 19)

The uncertainty and unreality of urban life, alongside intentions to return, reinforces group affiliations in town. 'Home' never loses its ideological significance, even for those who have never been there. In Blacksands though 'there is talk about "going home one day", many have never actually been "home", having been born and raised in town' (Mecartney 2001: 232). Children, and their education, declining language skills, the perceived need for money and the denigration of agriculture and fishing as means of livelihood have all kept people in town, even when they aspired to return and in the face of politicians' and other leaders' injunctions to 'go home'.

Throughout the Pacific there are also 'urban villages' of those whose once-rural villages have been engulfed by and incorporated in the new towns and cities. Such places have retained much of their social and political organisation, despite their economic livelihoods being utterly transformed. Their affiliations are to each other and to nearby rural kin. Despite urban employment, they are often antagonistic to the city and to its other residents, especially where land rights are threatened; yet, because of prior occupation, they are often powerful through numbers and the legitimacy of their claims to special treatment. Urban growth and some forms of urban management are threats, not promises.

Social groups in the cities are necessarily different from those in the home areas, since numbers in the city may be small and certain categories of kin likely to be absent. Urban households are less likely to be made up of a 'complementary labour pool', as kinship ties are weaker, hence friction within households is more common (Monsell-Davis 2000: 211–12). Nevertheless social affiliations do form and are developed in the manner that they would be in rural areas. There are hierarchies of such affiliations. These local connections in turn form the nucleus of social clubs recruited on the basis of their 'inland', 'coastal' or 'small island' origin (Battaglia 1986: 8). In the particular case of urban PNG not only do gangs form on the basis of home and ethnic origins, but rural tribal warfare may accentuate urban cleavages and spill over into similar military action in town (M. Ward 2000). Though friendship ties may be stronger than clan ties (Strathern 1975: 170), and workmates more important than kin, these exceptions enable the majority to follow the norm and stress its ideological significance. Friends are kin; others are strangers and perhaps even enemies.

In the largest cities encounters with others are transient, being characterised by standardised workplace and market interactions and myriads of temporary commercial transactions but always in the context of difference: especially ethnic variations, again most striking within Melanesia, and the opaqueness of linguistic variation. Encounters with those who are socially peripheral – islanders from other cultural groups – are fraught with uncertainty, and thus:

> charged with excitement, stemming from a sense of freedom, potentiality and adventure. . . . Although liaisons [between young men and women] occasionally occurred, more typically the efforts by a young man to impress a young woman from another group provoked a fight between the young men of both groups.
>
> (Gewertz and Errington 1991: 104)

Such disputes may often evolve into extended tribal fights within town, as occurred in Lae late in 1992, though they may merely replicate rural divisions. Numerically small urban cultural groups, such as the Trobrianders in Port Moresby or even Niueans in Apia (after a century), feel vulnerable; the former explicitly try to ensure that competition is intra-ethnic, and express concern that others might misinterpret their self-esteem as defiance and give them trouble on the streets or football pitches (Battaglia 1986: 20–1). Insecurity inevitably reinforces identity, most evident in the efforts of Gorokan migrants in Port Moresby to express their own rural village identity by violence against other Gorokan migrants (M. Ward 2000: 234–5), so inverting the strength of regional affiliations. At the same time connections with others are generally considered to be an advantage and prestigious, but difficult to obtain, retain and place trust in. In the smaller towns the formalities of culture contact are more apparent and less anonymous – and thus less fraught with tension – as in Wewak:

> Although there was ordinarily only minimal interaction between these Papua New Guineans, such as an occasional greeting in Pidgin English, most knew each other, often by name, and certainly by village or region of origin and by area of residence.
>
> (Gewertz and Errington 1991: 103)

Encounters between Pacific islanders and Europeans (and other migrant groups) are almost entirely generated in the context of employment and commerce; in this there is continuity rather than disjuncture. The external world is most likely to be perceived geographically, but that geography is contingent on ethnicity and highly flexible.

Informal squatter settlements increasingly characterise city life. Over time migrant cultural groups in towns may have therefore become more rather than less concentrated, though for any single group such as Trobrianders there is considerable dispersion (Strathern 1975; Battaglia 1986) as there is for even small migrant groups in formal housing contexts (Connell 1988a). Moreover, urban squatter settlements have sometimes come to resemble rural villages; as kinship ties in towns become more elaborate, residential patterns replicate village life, trees and gardens reach fruition, rituals are enacted in an urban setting, incomes are directed to local social objectives, urban leaders emerge and 'village' courts provide social control (see, for example, Goddard 1998). Nevertheless urban Trobrianders, 'if not isolated, are at least at sea when it comes to achieving the sense of collective physical boundaries and definition that hamlets and villages provide at home' (Battaglia 1986: 10). In dispersion they have developed their traditional harvest festivals on the unpropitious dusty slopes of Port Moresby, complete with the magical stones and sexual taboos that should ensure fertility: 'the satisfaction people speak of in clustering yams into gardens is a response to the scatteredness of their own living arrangements' (ibid.). Enacting such rituals provides a moral dimension in opposition to the dissoluteness of urban life. The Trobrianders, like so many others in town (see, for example, Lieber 1977; Flinn 1990; Zimmer-Tamakoshi 1998), combine nostalgia, ideology and alienation into the retention (and modification) of custom and the reassertion of identity.

More permanent urban residence, far from paradoxically, allows, even ensures, such efflorescence of culture and identity. Identity is reinforced by the differences readily apparent in urban life and is dependent on what it excludes. Culture and ethnicity are the most essential components of personal identity, most obvious in the difficulties that are posed in marriages across cultural boundaries (Strathern 1975; Connell 1988a). In the cities therefore social and economic relationships are largely based around a local place (a 'settlement') or kinship relations that link urban and rural places (rather than on personal relationships of friendship or workplace ties). Religious cosmologies, both pre-modern and recent, and tradition tend to be past-orientated; leadership within settlements may invoke religious beliefs (or sorcery) rather than political authority (cf. Giddens 1992: 100–6). Urban islanders, like the Trobrianders in Port Moresby, define themselves through their cultural identity in the self-esteem and pride that follows the appropriate and successful attention to tradition (however this may have been reconstructed in town).

Growing security concerns have provided a contrary differentiation of the city, notably in Port Moresby. Europeans and other expatriates have increasingly withdrawn to the confines of such hillsides as Tuaguba Hill and barricaded themselves behind high walls and fences, with guard dogs and elaborate security systems. Altitudinal differentiation, in diverse ways, has never been so apparent. The neo-colonial city has thus reverted to something akin to the more segregated colonial city: 'The identity of the modern city is created by what it keeps out. . . . In order to determine itself as the place of order, reason, propriety, cleanliness, civilisation and power, it must represent outside itself what is irrational, disordered, dirty, libidinous, barbarian and cowed' (Mitchell 1988: 165; Koczberski *et al.* 2001). Islanders may not be cowed, but such representations explain the elite distaste for the 'disorder' of squatter settlements (and opposition to them), in terms of the reluctance to condone their presence or provide them with services. In quite different ways security and tradition enable the replication of the old colonial order.

Urban life may be more exciting than a rural existence but can also be hard and difficult. Urban residents, above all those in the squatter settlements, often come from relatively marginal areas where income-earning opportunities are few and education is limited, hence they are, therefore, poorly equipped with resources, skills, power and connections to gain employment and cope with urban life. City incomes are often barely adequate to support the migrants let alone provide for remittances to rural kin. Not only is urban life difficult, it may also be dangerous. Crime and gangs are more common in urban areas. Women, especially in Port Moresby but also in other Melanesian towns, are victims of a particularly gendered violence, where rape and domestic violence are not uncommon, accentuated in cities, where men perceive they have lost control over women – an embattled masculinity confronting modernity (Lateef 1990; Banks

2000). There is a widespread perception that urban life is violent and rural life relatively harmonious. This rhetorical view remains strong even amongst those with little or no experience of it (Ryan 1989: 22). Commitment to urban life is far less than absolute with this ideology powerfully in place, whilst ethnic identity is powerfully reinforced for those – particularly the children born in town – who have had no personal and direct experience of 'home' areas in their everyday life. Children are often sent to board with rural grandparents, especially when their linguistic skills are weak: in town rural life acquires new meaning. None the less, many without substantial incomes choose to remain in town, despite the pervasive nostalgia for rural life. In the case of long-term Chambri migrants in Wewak, they chose 'freedom': escape from big-men (and their perceived arbitrary social control) and from sorcerers. Yet even where urban residents seek and prefer permanence, they do not discount or ignore their cultural identity (Connell 1988a; Ryan 1989; Jourdan 1994; Gewertz and Errington 1999: 116–17) nor are they able, or allowed, to choose to do so.

Some urban islanders have sought to disavow rural life, ethnic identity and cultural commitment and have adopted urban life styles. Such people, and households, are those who perceive rural commitment as irksome and unnecessary in an era of modernisation. They stress that rural folk should work hard, be self-reliant and not wait for the government or kin to help them. They are invariably those who have succeeded in a career (Connell and Lea 1994: 277–8; Gewertz and Errington 1999). By contrast, gang members in Port Moresby are part of another urban group, being mainly those who were born and raised in the city and thus said to be

> largely alienated from their 'home' areas and customs. Thus, while they may maintain the ethnic sensibility which characterises Papua New Guinea society in general, they share a transcendent urban lifestyle and world view, which . . . overrides the divisions which commonly exist between ethnic groups.
>
> (Goddard 1992: 23)

Still, beyond the gang members, some Melanesians are now retiring and choosing to stay in the city; a few are even being buried there. In Polynesia and Micronesia this is less novel.

For all their inability to differentiate themselves fully from the rural sector, some urban residents have nevertheless established a very different life style to that of their parents and many of their peers. They have kin, friends and colleagues. This 'apartness' is clearly recognised in the villages, where people see such life styles as attempts to put themselves above rural folk, a distinction symbolised in urban PNG through the pidgin term *susokman*, rather than the thong-wearing or barefooted villagers; those wearing socks and sunglasses are assumed to have ready cash (Strathern 1975: 233). An urban elite choose to distance themselves behind darkened windscreens in air-conditioned cars, sure targets for criminals, envious of and unimpressed by such inequity and alienation. Such elites, including company managers, public servants and politicians, self-defined through their consumption and participation in such activities as Rotary and golf clubs, and sometimes of mixed-race origins, fear the 'levelling demands' of villages were they to return there; they occupy particular suburbs and are becoming an incipient class (Gewertz and Errington 1999: 142). Extreme differences within Melanesian cities occur elsewhere in more subtle form.

Differences within cities, again especially in Melanesia, are sharpened and enhanced by the lack of national identities. Nation building has proved difficult and, in PNG at least, even efforts to agree upon a national flag, anthem, language or the name of the country have created dissent and division. In countries of considerable physical but, above all, ethnic and cultural diversity, where contact, development and decolonisation have all come late, the task of creating a nation and a national identity is enormous. Even the circumstances of early urban life slowed nationalism; for Melanesians 'the texture of their urban existence must surely be one of the most important reasons for the slow development of nationalism among a people who were never allowed to mix freely together away from home' (Wolfers 1975: 50). More recently the design of national symbols – the parliament houses – of the new states has challenged architects because of their importance in contexts where Melanesian cities lack corporate municipal institutions, and especially the imposing public buildings that might contain an institution and represent it (Mitchell 1988: 59). Moreover, even what passed for a semblance of colonial planning has been overwhelmed by the infilling and largely unforeseen urban consolidation of squatter settlements in the marginal sites that the formal sector ignored. The symbols of national identity have been difficult to achieve, and cities have scarcely been the crucibles of nationhood.

Cities apart?

If Pacific island cities are intricately subdivided, with ethnicity, regional and cultural identity prevailing over class divisions, they are assuredly part of a post-colonial world that has contributed to urbanisation and, as yet, to hesitant and undefined movements towards the formation of national and regional cultures and identities. They are nevertheless also part of the world system. The economies of the Pacific island states, but especially of the cities – the entrepôts and administrative centres – depend on fluctuations in the world capitalist system. Notwithstanding the continued significance of subsistence economies, trade systems and price fluctuations (that cannot be influenced by the Pacific's marginal presence) have embedded towns, cities and villages in a global economic, cultural, technological and political system, though there is scope for endless debate on the extent of globalisation in every sphere and place.

The island Pacific, its towns and cities, are part – though very much a peripheral part – of the world capitalist economy. Indeed, despite the recency of modern economic development, the dominance and significance of the world economy is particularly apparent because of the limited importance in the region of the state, itself a function of newly won independence (and lack of nationalism), and the clamour for foreign investment. Foreign investment is concentrated in commerce, tourism or resource-based activities like mining and forestry. There is minimal internal economic hegemony and only rare attempts to flex national if not local muscles against multinational corporations. All these characteristics are particularly striking in countries where many live without key benefits of modern technology.

Throughout the region modernity has resulted in a disjuncture between space and place. In pre-modern times space and place largely coincided, since the spatial dimensions of social life were for most dominated by localised activities. External links were tenuous and often highly dangerous, epitomised by the 'barter markets' where 'traders laid down their goods and withdrew to a safe distance while the exchange was made' (Gewertz 1978). Subsequently, economic relationships, notably commodity pricing in one form or another, have influenced local life for almost every Pacific society. Stores have become the final points of distribution in the world system; their goods with their distant origins, are the symbols and substance of the diverse structure of incorporation and accommodation (McInnes and Connell 1988; Curry 1999). In a sense place has become 'increasingly phantasmagoric' (Giddens 1992: 19) as localities are thoroughly penetrated and influenced by very distinct social, political and economic influences, whose genesis and local ramifications are often beyond comprehension.

Culture and identity are not direct products of new, increasingly global economic (and perhaps political) arrangements – and their socio-spatial impact on the material form and social structure of cities (Smith 1992) – but are shaped by the everyday practices of ordinary people and their feelings and understanding of their conditions of existence. Globalisation of social and economic life is far from complete; there are many people, especially in Melanesia and Micronesia, who are very distant both in space and impact from the institutional transformations of modernity, whilst there are reactions against, and responses to, modernity. The processes of development have mapped out a contested landscape of environmental, economic and cultural change, but all cultures are products of a history of appropriation, resistance and accommodation. In the Pacific struggles for transformation have taken diverse forms, but few are urban, women's, ecological or, in a broad sense, concerned with human rights (cf. Slater 1992: 311). Indeed, the complex recent struggles in Bougainville, Solomon Islands and Fiji epitomise Appadurai's (1990: 304) general notion that ethnicity has become a powerful global force and, here as elsewhere, 'state and nation are at each other's throats'. Local struggles of resistance are more important than either regional, class or gender issues. At the same time 'contemporary communications have collapsed the "usual space and time boundaries" and produced both a new internationalism and internal differentiations within cities and societies based on place, function and social interest' (Harvey 1989: 75) but such 'usual' space and time boundaries were never immutable. Social relationships were invariably highly flexible to accommodate shifting allegiances, opportunities and domestic events, including natural disasters: 'polyphonic patterns of accommodation and resistance to domination' (Smith 1992: 496). Despite various forms of resistance and accommodation, Pacific cities – and the rural hinterlands beyond – are very much part of, and not apart from, world systems. Whilst their diversity is unusual, these cities of parts are increasingly being shaped by forces emanating beyond national boundaries, whilst nascent attempts to plan and manage such cities have similarly evoked global parallels and policies.

The management of disarray and diversity

One outcome of the changeover from colony to independent statehood in the 1960s and 1970s was a rapid weakening in the inherited mechanisms which governed urban development. Western urban planning systems are both prescriptive and proscriptive and depend ultimately on popular acceptance of government intervention in everyday life. Such interference is tolerated because local government is usually reasonably responsive to local wishes and its powers are clearly defined. Urban management is one facet of a complete system of local governance and alterations to some of its parts affect the whole. Even where they exist, municipal governments in small developing countries have rarely enjoyed the preconditions necessary for them to operate effectively. They are commonly subjugated by the national government located in the same city and make do with scant resources. Some towns simply have no urban governments at all. There are therefore no authorities with specific responsibility for managing urban affairs, which are therefore dealt with in the broader context of national development. Nor are there enough suitable people able and willing to stand for election, given the alternative opportunities open to such people in the public and private sectors and national government, or to participate in CBOs. Post-colonial change in municipal affairs was marked by the declining role of regulations, with reduced attention paid to land use zoning and building by-laws, but also by the failure to maintain and provide for the major urban infrastructure of roads, water supplies and sewage treatment. This was not due entirely to changed priorities, a shortage of funds for urban investment or the adoption of new principles, but was rather the outcome of less adequate management and the ignoring of regulations created to manage urban affairs in a different political context.

Throughout the region the dialogue between democratic political process and local prestige and authority is barely beginning, even in urban areas. Any tradition of the 'common good' in a national sense is of very recent origin and more honoured in the breach, whilst governmental infrastructure and legal precedents are not yet developed. Moreover, in many countries, there has been a shift in power and administrative authority from public servants to politicians, against a background of worsening economic conditions. As government officials have lost power to politicians, long-term economic and physical planning and resource management have been abandoned in favour of short-term expediency (Frazer 1997: 47). Despite the emergence of NGOs, some with a focus on environmental issues, local participation in all facets of development and planning has failed to emerge. Local people are poorly informed, even on matters of direct concern to them (see, for example, Atoll Research Programme 1996), but are also no less interested than politicians in short-term solutions and particularly absorbed with issues that directly affect their own local situation (Adams 1982; Mecartney 2001). Regional and provincial governments have been no more successful (Ward 1990). Moreover, some of the symptoms of poor governance – the failure to make a clear distinction between public and private interests (hence the tendency to divert public resources to private benefit), the failure to establish a predictable framework for law and government behaviour conducive to development (rather than arbitrary application of rules and laws) and excessive regulation and licensing requirements (involving misallocation of resources, and non-transparent decision-making) (Ray 1998: 3) – have all hindered development and urban management. Although urban communities are sensitive to falling service standards and declining infrastructure provision, there is residual belief in the need for government to take action and formulate new policies when problems may be more fundamental and point to the need for changes in the structure of government itself. Indeed, current social and political systems may have more support than potentially fairer yet alien systems (Storey 1998b: 74). Throughout the region public participation has been scarcely tried and rarely welcomed.

Urban councils, where they exist at all, are characteristically weak, uninformed and with few powers, whilst traditional leaders in urban areas lack the authority of their rural counterparts. Practical change in many towns is stimulated by the national politicians who represent these areas. Such politicians have one key objective: re-election. Local groups may be rewarded with cash donations for such projects as water tanks and improved roads; such donations are cyclical, linked to election times rather than to development plans. In the 1997 election campaign in PNG, the Prime Minister promised one settlement, composed mainly of migrants from Morobe, a surfaced basketball court, sealed roads, electricity, a sewerage system and properly surveyed housing blocks. None eventuated (Barber 1999). Some politicians have actually established new settlements, with promises of land ownership and utilities, in reward for re-election. In Port Vila, it is widely believed that the present Prime Minister, Barak Sope, a local landowner, has placed settlers in particular areas on condition that they vote for him (Mecartney 2001). This emphasises the fragility of tenure and the insecurity of urban residence, creates an unwillingness to participate in political issues of any kind, and challenges any notion of development planning. In Blacksands, where establishing local organisations across a community of considerable social diversity had proved difficult though there were island-origin and church groups, 'When asked what the government could do to help, fully 100 per cent of the responses were loud snorts of laughter' (ibid.: 225). Settlers were acutely aware that their only chance of assistance came briefly at election time. Moreover, there as elsewhere, to develop effective ties within a local community, let alone be proactive, is beyond the capacity of settlers who are relatively impoverished and limited in knowledge of the structures of management, in a context where neither landowners nor governments are willing to provide assistance to them.

Traditional approaches to problems of urban management (or, more generally, to social and economic development) are of limited value both because of the scale at which such approaches operate, the often hierarchical nature of indigenous societies (and thus the marginalisation of particular groups) and the diverse aspirations and interests of people within small-scale societies. Social cohesion is always fragile, and pre-colonial societies were neither egalitarian nor communal; their descendants are not so now. It was nevertheless anticipated that the abandonment of colonial institutions and their replacement by more indigenous forms of governance would solve the problems and injustices of the colonial period. That failed to eventuate, as did the lack of administrative capacity in relation to the rational ideas of the bureaucratic system, with

> the decision of new Pacific island states to use the public sector as a means of job creation and wealth distribution [hence] the efficiency and effectiveness of the public sector, as much as it ever existed, declined. . . . Because government is weak and incompetent, people have no faith in it, their cynicism and recourse to alternative means of self-help keeps government weak and incompetent.
>
> (Schoeffel 1997: 8)

Planning and management fell between two stools; neither traditional nor modern concepts of these activities were effective. Indeed, whilst the majority of Pacific populations remained rural, and many urban residents were sojourners not settlers, urban planning could never have been more than embryonic.

Urban heritage is consequently being made on the run. A distinctive form of urbanisation has appeared around the rights or non-rights of urban residents to customary land. Cities are characterised by rapidly growing uncontrolled fringes, settlements on marginal land beyond the reaches of the formal housing sector and pockets of traditional villages now swallowed up in the expanding modern town. Modern banks and offices, new tourist establishments and the expensive dwellings of the elite (still partly expatriate in places) coexist with low-income suburbs and place huge demands on poorly developed networks of infrastructure services. This complex and highly differentiated urban mosaic is rarely under the jurisdiction of a single municipal authority and is the chief theatre of operations for a host of national government agencies, NGOs, private sector interests and landowning groups. Sustained maintenance and management of urban services appears an unrealistic dream.

The general lack of an integrated context for economic and physical planning, even a basic land use plan, has created difficulties in the provision of urban services. Planning and service provision have been managed by a range of different organisations with different agendas, principles and policies and without real coordination or cooperation between them. While there are costs involved in establishing an appropriate framework for adequate planning and management, these are likely to be much less than those associated with rapid urban growth in an unplanned context. Though several countries in Asia have made attempts to introduce integrated urban infrastructure development programming (IUIDP) to tackle urban problems systematically, even there, with the greater resources of national programmes supported by international agencies such as the World Bank, the institutional complexities involved in service delivery have meant that the introduction of IUIDP was limited in the first instance to particular areas like water supply and housing. None the less the general concepts supporting IUIDP procedures (sustained economic productivity, basic needs fulfilment, integration of plans, programmes and projects, and decentralisation of decision-making) are of relevance in the Pacific. Orderly investment programming for urban services delivery rests on the clear identification of infrastructure priorities. But it also rests on sustained economic growth and, in its absence, a critical dependence on overseas aid.

Effective planning presumes an ability to decide for the collective good of all and manage the outcomes. This has proved very difficult in small countries where there is scarcely room for two levels of government in the capital city; the result is domination by central government and its consequent and often inappropriate involvement in municipal affairs. The future of urban management is very much bound up with the ability of urban authorities to become more self-sufficient. The potential for achieving this depends on a complex set of factors at the national, regional and local levels. Sources of revenue are taxes (licences), fees (for service provision), and grants, but taxes and fees are hard to collect, grants difficult to obtain and few urban areas are even legislated to undertake this. Solutions consequently materialise in episodic and expensive responses to crisis conditions, without reference to the wider context of urban service provision. Urban management has become crisis management, rather than the good housekeeping that might avert the worst crises, and emphasises grandiose 'solutions' rather than planned development.

The same problems of mismanagement and institutional paralysis are apparent for service provision. Despite the international priority given to the supply of adequate water and sanitation facilities, there is little evidence that these and other basic needs have any real priority, other than at times of breakdown. Less than adequate investment characterises service provision in many key areas: water supplies, sewerage facilities and environmental management (street and stream cleaning and garbage disposal) and also in the provision of open spaces, transport services, health centres, roads and electricity. There is however a very strong correlation between access to formal housing and access to other urban services. A common characteristic in developing countries is that the best water and sanitation services are provided to the more affluent areas of cities even though these may be more expensive to service from an engineering perspective (Ludwig and Browder 1992), since they are often on hilltops, whilst poorer areas are correspondingly supplied inadequately or intermittently:

> The city authorities of Port Moresby have to seriously address the future of a safe reliable water supply to the city. . . . The water system is now over thirty years old and can no longer cope with the increasing demands. . . . The city's excuse about old water pipes and not enough pressure to fill up the reservoirs is beginning to wear thin now. This is becoming apparent by the senseless acts of vandalism that occur to the pipes each time water supply to parts of the city is disrupted for longer than a day . . . it is time to stop and start some serious constructive planning for the future.
>
> (*TPNG*, 16 July 1992)

Such plaintive editorials, alongside the observations formalised in consultancy reports, are familiar throughout the region. In situations of reduced water supply, frequent in one of the driest cities in Melanesia, the tower blocks of the affluent Port Moresby population metaphorically drain scarce water supplies upwards and away from the poor. This deterioriating and inequitable situation is broadly true elsewhere, as the public works systems installed in colonial times creak and rust into obsolescence. Leaks, storm overflows, shortages in dry weather, ageing and damaged assets, illegal connections and failures to collect revenue have all exacerbated the basic problems. Colonial order – albeit more inequitable – has not survived.

Throughout the Pacific there are problems of access to housing and both social and physical variations in character within towns are in part determined by, and in part reflected in, the availability and provision of housing. Within the towns and cities there are enormous differences in residential standards. In Port Moresby, for example, the extremely expensive homes and multistorey luxury flats of Tuaguba Hill, overlook low-cost NHC buildings and the flimsy wooden shanty towns. Everywhere 'marginal' housing reflects status: illegal or barely tolerated and hence somewhat insecure. Yet even between settlements there are enormous differences; the provision of finance for housing is a political issue, funds are scarce and some settlements are populated by groups who are believed to be 'trouble makers' and criminals, hence their claims on finance for residential improvement are less likely to be met. Scarcity of finance and therefore of housing (and related services) reflects and often entrenches urban inequalities. At best, recent policies towards settlements can be described as barely benign neglect. Institutional paralysis has stimulated collaboration and self-help, both legal and illegal, within the settlements. A testimony to the resourcefulness of the poor in Pacific cities is that they have found ways of housing themselves which are appropriate, flexible and of a higher standard than in some other parts of the world.

Urbanisation comprises two key demographic components – relatively high rates of population growth as well as increasing concentrations in cities – which are not synonymous. In policy terms the former suggests population policies whereas the latter points towards decentralisation and improvements in rural development (Slater 1986). Both imply a third focus on urban policy, including the strengthening of municipal government and the institutions governing the delivery of urban services. None of these three areas has resulted in effective policy formation in the Pacific, despite myriad rhetorical statements. Governments have generally lost interest in and withdrawn their funding from urban development as the challenges mount. Globally, one urban analyst has concluded:

> My pessimistic prediction is that by about 1995, government regulations will have been cut back a great deal and people will have been left even more to their own devices. No doubt this will have brought some benefits, especially for the affluent minority who can afford to pay for quality services. But the majority will have discovered how exploitative most illegal subdividers are, about the dangerous state in which most private companies run their buses, and how few private entrepreneurs are prepared to provide services to the unprofitable poor.
>
> (Gilbert 1992: 436)

In Oceania this message has already come true. One outcome is greater inequality whilst further attempts are being made to divest government responsibilities for some essential services. Whether this is an admission of defeat in the face of a mounting crisis, sheer bureaucratic inertia or a belief that some services cannot be adequately managed by government varies but the effects are far reaching. Moreover, urban management is now more difficult under SAPs, and, as in African towns (Briggs and Yeboah 2001), has resulted in expanded settlements rather than decentralisation, a situation that has made towns more like urban areas (through increased agriculture, informal economic activity and self-help) though the necessary 'urbanisation' of the rural areas, via improved access to services and jobs, has yet to occur.

Although there were early attempts to formulate some coherent urban policy throughout the region, present policies towards the management of urbanisation generally consist of piecemeal plans and projects in particular towns. Coordination is conspicuous by its absence. This is a direct reflection of the condition of urban government across the region and is unlikely to change until the structure of urban governance changes. What was said long ago about planning for urbanisation in developing countries in general is strikingly true of the Pacific:

> What characterises most of the planning efforts in the third world is the absence of a will to plan effectively. Most urbanisation policy is unconscious, partial, uncoordinated and negative. It is unconscious in the sense that those who effect it are largely unaware of its proportions and features. It is partial in that few points at which governments might act to manage urbanisation and affect its course and direction are in fact utilised. It is

> uncoordinated in that national planning tends to be economic and urban planning tends to be physical and disjunction often produces competing policies. It is negative in that the ideological perspective of the planners leads them to try to divert, retard or stop urban growth and in particular to inhibit the expansion of metropolitan and primate cities.
>
> (Berry 1976: 78)

In the Pacific the disjuncture between economic and physical planning has substantially increased and become formalised. Current policies are largely aimed at economic growth, rather than more broadly based economic development, and are unlikely to reduce problems associated with equity and the distribution of urban (and other) services. Fragmentation, and tension between landowners and migrants in the face of land shortages, growing urban unemployment, bureaucratic ineptitude and political corruption – hastening privatisation – have all contributed to disarray and division. Most aspects of infrastructure provision can only be addressed with the support of the urban population at large. In the Pacific, history, geography and the gulf between 'traditional' and 'modern' has made the creation of this 'community of interest', or any notion of it, very difficult indeed. Diversity and spontaneity are more apparent.

Achieving sustainable urban development is threatened by a multitude of factors, in a region where even sustainable agricultural societies were more myth than reality (Connell 2001) and medium-term planning is absent. Private sectors are more powerful than public sectors – even, as in Micronesia, where the former is merely characterised by a wealth of stores and taxis – and largely beyond regulation. Basic planning functions are absent, weak or fragmented, and sectors are far from complementary, hence infrastructure is extremely costly (especially where economies of scale are absent, and key components are imported), privatised utility corporations are underfunded, poorly staffed and intent on profitability rather than equity. But at the core of such technical, economic and political issues – characteristically described as a weakness of political will – are fragmented communities of sometimes temporary townsfolk, competitive rather than cooperative, with local or rural rather than urban interests, sometimes poorly informed, rarely consulted and usually ignored (except at election times). For historic residents of urban centres, whose lives are 'written on the ground', anxieties over land – their sole strategic resource – have resulted in tensions over development, resistance to alienation and costly, fractured and fragmented urban expansion. Acute social divisions are embedded within poorly performing urban (and national) economies, where rhetoric champions rural development (and return migration) and reality hinders it, through urban bias in resource allocation (see, for example, Connell 1986). Island states are not wealthy and there are competing priorities for income expenditure. The tenets of sustainable development, whether social (quality of life), political (good governance), economic (generating economic growth) or environmental, and all are part, involve considerations of local and global communities and the rights of future populations, and are challenged by hierarchy and ethnicity (and emerging classes), weak capital formation, and shifts to the 'right' through privatisation, deregulation and corporatisation, where inadequate financial resources, exceptionally short-term political horizons (especially in Melanesia, where frequent parliamentary votes of no confidence deter anything other than survival) and limited human resources ensure that ethics are subsumed under self-interest, immediacy or inertia. Set within a contemporary framework of structural adjustment, deregulation and cost recovery, with the downsizing (euphemistically, in Vanuatu at least, 'right-sizing') of bureaucracies, governments have become both less willing and less able to assume planning responsibilities, whilst urban residents are no better informed or organised, in what is often an information vacuum. Sustainable development may be as distant now as it has ever been.

Future cities

The 'last' parts of rural Melanesia were probably not 'contacted' (an imprecise term) until the 1970s – in the final flurry of world exploration – yet throughout the region universal images, sounds and brand-names now abound: T-shirts (even Gucci and Benetton look-alikes), Toyotas, Coca Cola, sunglasses, digital watches and mobile phones, and a host of video stores. In this universalisation the Pacific has gone from pre-literacy to post-literacy, with only a rudimentary intervening stage. Though the urban elite have ensured a degree of distinctiveness, with shirts, ties and offices, eating 'takeaway' Chinese meals and Kentucky Fried Chicken and tuning in to overseas television channels, they do not decorate their homes with indigenous artefacts – as Europeans do – but play golf and may travel more frequently outside the country than to their 'homes', whose people and interests they so often seek to represent and benefit from. 'Ethnic and cultural fragmentation and modernist homogenization are not two arguments, two opposing views of what is happening in the world today, but two constitutive trends of global reality' (Friedman 1990: 311). The urban juxtaposition of corporate monoliths and plywood shacks, air-conditioned vehicles and bare feet is not post-modernist but universal and, in different formats, long-established between rich and poor: it is the era of late modernity in impoverished states of the periphery. On a different scale, for a different audience, images of 'tradition' are transmitted globally – through tourism, travelogue and the splendours of tropical nature – whilst global culture in diverse forms flows through the satellite dishes of the cities. 'Tradition' moves out as 'modernity' moves in.

Consumption demarcates incipient class divisions, yet ethnicity and also gender complicate and minimise emergent boundaries within ethnic groups. Women, more confined – especially as urban crime increases – are more conscious of identity. Yet in the Pacific, as elsewhere:

> There is relatively little consideration of how people define themselves, how identities are cobbled together to act in the new spaces of a post-fordist

> economy. Amidst the shards of modernist fragmentation and disengagement, how can identity be constructed at all?
>
> (Watts 1992: 123)

In states that are scarcely nations, where tribes are also recent inventions – local and national constructs – and urbanisation has preceded industrialisation, the task becomes more difficult. Yet, as with the Chambri of PNG, few question who they are. Linguistic and philosophical identity, even of such long-established urban residents as the Toaripi (Ryan 1989), stimulated memories of land tenure, where land is owned by the ancestors and held in trust by the living for those as yet unborn, are as important as the recent trappings of modernity.

Elite islanders, especially those who have married outside their immediate cultural group, have tended to become a more permanent urban population. This is particularly true of bureaucrats – with long employment tenure – who, if not already in capital cities, seek to migrate there. Similarly, most university students seek to, and do, remain in the capital (Weinand and Ward 1979). Of the bureaucrats especially:

> Their presence in urban areas not only deprives rural areas of what skills they possess, but ensures a continuing urban bias to the economy through their greater purchasing power and their ability to influence decision-making. The children of urban-dwelling public servants are the first of a truly urban generation of Papua New Guineans.
>
> (Allen 1983: 228)

Yet there are few of them, in PNG or elsewhere, and their urban identities are weak. There are distinct attractions to urban life for some and though these attractions provide personal gains they do not challenge the ideological construction of ethnic identity. Urban migrants do not see their ethnic identity as diminished by development, education or monetary success; rather they incorporate these elements into their social and political life. They remain in opposition to other ethnic groups, through competition for land, political influence and economic influence.

To retain urban privileges urban resources must be diverted from the poor, hence policies that favour equity in service distribution have faded from view (or are merely not implemented) while strong opposition is expressed to the growth of poor, migrant, urban populations (Chapter 3). A rural ideology has held very firm in the minds of long-established urban residents but especially where their urban privileges have been at stake. Evident in the opposition to urban squatter settlements is the manner in which historic European colonial perceptions have become replicated in contemporary Pacific perceptions. In a sense the colonial order has penetrated and colonised local discourse (Denoon 1980; Mitchell 1988; Koczberski *et al.* 2001). The elite of the post-colonial era have adopted some of the mores of the colonialists, so reinforcing old divisions and antagonisms within the cities.

Pacific urbanism emphasises the role of historical specificity, plurality and difference; ultimately there are few universals in a region where culture (and emotion) are of pervasive importance. Local languages and the (usually geographically limited) circuits of communication in which they are involved are resources as well as limits, partly acting as shields against 'global' influences (Seers 1983: 69–76; Smith 1992: 496–7), however these may be defined in a world where local and global are imprecise. Because the Pacific – since long before colonialism – contained an astonishing variety of languages, cultures and ecologies, identities are fluid, and dissent and conflict have produced complex cities and nations. Notions of ethics, rationality and justice are diverse. Political and bureaucratic practice barely yet puts identities at risk, hence the tasks of urban planning and management have proved unusually difficult, and scarcely take account of, or provide a voice for, the different groups who live in the cities without forming wider communities. Where land retains its crucial role, and as Pacific cities move into a new millennium, urban life will continue to display extraordinary diversity and incoherence.

References

Adams, R. (1982) 'City and village: the Ahi response', in R.J. May (ed.) *Micro-nationalist Movements in Papua New Guinea*, Canberra: Australian National University Political and Social Change Monograph no. 1, 207–45.

Agyei, W. (1988) *Fertility and Family Planning in the Third World*, London: Croom Helm.

Ahlburg, D. (1991) *Remittances and Their Impact: A Study of Tonga and Western Samoa*, Canberra: Australian National University, National Centre for Development Studies, Pacific Policy Paper 7.

—— (1992) 'Fertility and family planning in Papua New Guinea', unpublished report, Minneapolis.

—— (1996) *Demographic and Social Change in the Island Nations of the Pacific*, Honolulu: East-West Center, Asia-Pacific Population Research Report no. 7.

—— and Levin, M. (1990) *The North East Passage: A Study of Pacific Islander Migration to American Samoa and the United States*, Canberra: Australian National University, National Centre for Development Studies, Pacific Research Monograph no. 23.

Alexander, W. (1978) 'Wage labor, urbanization and culture change in the Marshall Islands: the Ebeye case', unpublished PhD thesis, New School for Social Research, New York.

Alkire, W.H. (1993) 'Madrich: outer islanders on Yap', *Pacific Studies* 16: 31–66.

Allan, C.H. (1957) *Customary Land Tenure in the British Solomon Islands Protectorate*, Honiara: Western Pacific High Commission.

Allen, B.J. (1976) 'Maprik', in R.T. Jackson (ed.) *An Introduction to the Urban Geography of Papua New Guinea*, Port Moresby: University of Papua New Guinea, Department of Geography, Occasional Paper no. 13, 336–55.

—— (1982) 'Pre-contact trade in Papua New Guinea', in R.J. May and H. Nelson (eds) *Melanesia: Beyond Diversity*, Canberra: Australian National University, Research School of Pacific Studies, 193–206.

—— (1983) 'Paradise lost? Rural development in an export led economy: the case of Papua New Guinea', in D.A.M. Lea and D.P. Chaudhri (eds) *Rural Development and the State: Contradictions and Dilemmas in Developing Countries*, London: Methuen, 215–40.

Annez, P. and Friendly, A. (1996) 'Cities in the developing world', *Finance and Development* 33 (4): 12–14.

Anon. (1974) 'Moresby confounds the gloomy prophets', *Pacific Islands Monthly* 46 (6): 37–41.

—— (1999) 'Tightening the taps on water consumption', *Islands Business* 25 (9): 26.

Appadurai, A. (1990) 'Disjuncture and difference in the global cultural economy', *Theory, Culture and Society* 7: 295–310.

Arriens, W.L., Bird, J., Berkoff, J. and Mosely, P. (1996) *Towards Effective Water Policy in the Asian and Pacific Region*, vol. 1: *Overview of Issues and Recommendations*, Manila: Asian Development Bank.

Aruga, W. (1992) 'An overview of urban infrastructure services in Papua New Guinea', unpublished paper delivered to United Nations Seminar on Urban Information Systems, Beijing.

Asian Development Bank (ADB) (1992) *Papua New Guinea: Urban Sector Profile*, Manila: Asian Development Bank.

—— (1997) *Vanuatu: Economic Performance, Policy and Reform Issues*, Manila: Asian Development Bank.

—— (1998) *Urban Growth Management Strategy for Port Vila*, Manila: Asian Development Bank.

Atoll Research Programme (1996) *Community Environment Survey: Results of a Survey Conducted in Betio, Bairiki, Bikenibeu, South Tarawa, Kiribati*, Bairiki: Atoll Research Programme, University of the South Pacific.

Axline, A. (1989) *Financing Urban Local Government in Papua New Guinea*, Port Moresby: Institute of Public Affairs.

Bailey, E. (1984) 'Population distribution and migration', in *Report of the 1978 Census of Kiribati*, Tarawa: Government of Kiribati, 25–42.

Ballendorf, D. (1991) 'Money, action, and thinking: development planning in reverse sequence in Micronesia', unpublished paper to XVII Pacific Science Congress, Honolulu, Hawaii.

Bank of Hawaii (1995) *Federated States of Micronesia Economic Report*, Honolulu: Bank of Hawaii Economics Department.

Banks, C. (2000) 'Contextualising sexual violence: rape and carnal knowledge in Papua New Guinea', in S. Dinnen and A. Ley (eds) *Reflections on Violence in Melanesia*, Sydney: Hawkins Press, 83–104.

Banks, G. (1993) 'Mining multinationals and developing countries: theory and practice in Papua New Guinea', *Applied Geography* 13: 313–27.

Barber, K. (1999) 'Contemporary indigenous urban households in Papua New Guinea', unpublished mimeo, Port Moresby.

Barnabas, S. and Hezel, F.X. (1994) 'The changing Pohnpeian family', *Journal of the Pacific Society* 17 (2–3): 15–26.

Barr, K.J. (1990) *Poverty in Fiji*, Suva: Fiji Forum for Peace, Justice and Integrity of Creation.

Bartelmus, P. (1992) *Sustainable Economic Growth and Development: Concepts and Strategies*, New York: United Nations Department of Economic and Social Development, Working Paper No. 18.

Bathgate, M. (1978) 'Marketing distribution systems in the Solomon Islands: the supply of food in Honiara', in P. Rimmer, D. Drakakis-Smith and T.G. McGee (eds) *Food, Shelter and Transport in Southeast Asia and the Pacific*, Canberra: Australian National University, Department of Human Geography, 63–100.

Battaglia, D. (1986) *Bringing Home to Moresby: Urban Gardening and Ethnic Pride among Trobriand Islanders in the National Capital*, Port Moresby: Institute of Applied Social and Economic Research Special Publication no. 11.

Bayliss-Smith, T., Bedford, R., Brookfield, H. and Latham, M. (1988) *Islands, Islanders and the World: The Colonial and Post-colonial Experience of Eastern Fiji*, Cambridge: Cambridge University Press.

Bedford, R.D. (1968) *Urban Tarawa: Problems and Prospects*, Auckland: University of Auckland, Department of Geography.

—— (1971) 'Mobility in transition: an analysis of population movement in the New Hebrides', unpublished PhD thesis, Australian National University, Canberra.

—— (1973) *New Hebridean Mobility: A Study of Circular Migration*, Canberra: Australian National University, Department of Human Geography.

Bellam, M.E.P. (1964) 'The Melanesian in town', unpublished MA thesis, Victoria University, Wellington.

—— (1970) 'The colonial city: Honiara. A Pacific case study', *Pacific Viewpoint* 11: 66–96.

Belshaw, C.S. (1963) 'Pacific islands: towns and the theory of growth', in A. Spoehr (ed.) *Pacific Port Towns and Cities*, Honolulu: Bishop Museum Press, 17–24.

Belz, L.H. (1984) *Nuku'alofa: Sanitation and Reclamation*, Nuku'alofa: Ministry of Health.

Berger, B., Berger, B. and Kelner, H. (1974) *The Homeless Mind*, Harmondsworth, Middx: Penguin.

Berry, B.J.L. (1976) 'Comparative urbanisation strategies', in H. Swain and B. MacKinnon (eds) *Managing Urban Systems*, Laxenburg: Institute for Applied Systems Analysis, 66–79.

Bertram, G. and Watters, R.F. (1985) 'The MIRAB economy in South Pacific microstates', *Pacific Viewpoint* 26: 497–520.

Bonnemaison, J. (1974) 'Changements dans la vie rurale et mutations migratoires aux Nouvelles Hebrides', *Cahiers ORSTOM Series Sciences Humaines* 11: 259–86.

—— (1985) 'The tree and the canoe: roots and mobility in Vanuatu society', in M. Chapman (ed.) *Mobility and Identity in the Island Pacific*, *Pacific Viewpoint* 26: 30–62.

Boyd, D.J. (1990) 'New wealth and old power: circulation, remittance and the control of inequality in an Eastern Highlands community, Papua New Guinea', in J. Connell (ed.) *Migration and Development in the South Pacific*, Canberra: Australian National University, National Centre for Development Studies, Pacific Research Monograph no. 24, 97–106.

Briggs, J. and Yeboah, I. (2001) 'Structural adjustment and the contemporary sub-Saharan African city', *Area* 33: 18–26.

Briscoe, J. (1992) 'Poverty and water supply: how to move forward', *Finance and Development* 29: 16–19.

—— and de Ferranti, D. (1988) *Water for Rural Communities: Helping People to Help Themselves*, Washington, DC: World Bank.

Brookfield, H.C. (1980) 'The Fiji study: testing the MAB approach', in H.C. Brookfield (ed.) *Population–Environment Relations in Tropical Islands: The Case of Eastern Fiji*, Paris: UNESCO, 13–28.

—— and Brown Glick, P. (1969) *The People of Vila: Report of a Census of a Pacific Town*, Canberra: Australian National University, Department of Human Geography.

—— with Hart, D. (1971) *Melanesia*, London: Methuen.

Brown, R.P.C. and Connell, J. (1993) 'The global flea market: migration, remittances and the informal economy in Tonga', *Development and Change* 24: 611–47.

Bryant, J.J. (1977) 'Urbanization in Papua New Guinea: problems of access to housing and services', *Pacific Viewpoint* 18: 43–57.

—— (1990a) 'The acceptable face of self-help housing: subletting in Fiji squatter settlements – exploitation or survival strategy?', in D.W. Drakakis-Smith (ed.) *Economic Growth and Urbanization in Developing Areas*, London: Routledge, 171–95.

—— (1990b) 'Self-help in Fiji: squatter housing and informal employment as responses to poverty', in R.B. Potter and A.T. Salau (eds) *Cities and Development in the Third World*, London: Mansell, 86–99.

—— (1992) 'Poverty in Fiji: who are the urban poor?', *Singapore Journal of Tropical Geography* 13: 90–102.

—— (1993) *Urban Poverty and the Environment in the South Pacific*, Armidale, NSW: Department of Geography and Planning, University of New England.

Bryant-Tokalau, J.J. (1995) 'The myth exploded: urban poverty in the Pacific', *Environment and Urbanization* 7: 109–29.

Calvert-Faamoe, K., Quested, C. and Sio, B. (1997) 'Nutrition and Apia urban youth', in *Report on the Apia Urban Youth Survey*, vol. 4, Apia: Department of Statistics, 96–104.

Campbell, I. (1992) *Island Kingdom: Tonga, Ancient and Modern*, Christchurch: Canterbury University Press.

Chambers, A. (1986) *Reproduction in Nanumea: An Ethnography of Fertility and Birth*, Auckland: University of Auckland, Working Papers in Anthropology no. 72.

Chandra, R. (1979) 'Urbanization in Fiji, 1966–1976', unpublished paper to 14th Pacific Science Congress, Khabarovsk.

—— (1990) 'Patterns and processes of urbanization in Fiji', in R. Chandra and J. Bryant (eds) *Population of Fiji*, Noumea: South Pacific Commission, 157–79.

Chandra, S. (1977) 'Urban delinquency in Fiji', in J. Harré and C. Knapman (eds) *Living in Town*, Suva: University of the South Pacific, 57–68.

Chao, M. (1980) 'Socio-cultural change in Fiji', unpublished PhD thesis, Catholic University of America, Washington.

—— (1985) 'Life in a squatter settlement', *Catalyst* 15: 168–207.

Chapman, M. (1991) 'Pacific island movement and socio-economic change: metaphors of misunderstanding', *Population and Development Review* 17: 263–92.

Chen, T.Y. and Yala, C. (2000) 'The Malahang Industrial Centre and the development of industrial centres in Papua New Guinea', *Pacific Economic Bulletin* 15 (2): 59–74.

Christensen, P.M. (1995) *Infant Nutrition and Child Health on Tarawa, Kiribati*, Sydney: University of New South Wales, Pacific Studies Monograph no. 15.

Clarke, G. (1993a) *Life and Living in Tuvalu*, vol. I: *Findings, Report to the Government of Tuvalu*, Sydney: Australian International Development Assistance Board Pacific Regional Team.

—— (1993b) *Life and Living in Tuvalu*, vol. II: Strategies, Sydney: Australian International Development Assistance Board Pacific Regional Team.

Clifford, J. (1988) *The Predicament of Culture*, Cambridge, Mass.: Harvard University Press.

Cole, R.V. (ed.) (1993) *Pacific 2010: Challenging the Future*, Canberra: Australian National University, National Centre for Development Studies, Pacific Policy Paper no. 9.

Connell, J. (1973) 'Pacific urbanization: the exceptional case of Tarawa atoll', *Geographical Review* 65: 402–4.

—— (1976) 'Buin', in R.T. Jackson (ed.) *An Introduction to the Urban Geography of Papua New Guinea*, Port Moresby: University of Papua New Guinea, Department of Geography, Occasional Paper no. 13, 483–6.

—— (1983a) *Migration, Employment and Development in the South Pacific. Country Report no. 3: Federated States of Micronesia*, Noumea: South Pacific Commission and International Labour Organisation.

—— (1983b) *Migration, Employment and Development in the South Pacific. Country Report no. 7: Kiribati*, Noumea: South Pacific Commission and International Labour Organisation.

—— (1983c) *Migration, Employment and Development in the South Pacific. Country Report no. 8: Marshall Islands*, Noumea: South Pacific Commission and International Labour Organisation.

—— (1983d) *Migration, Employment and Development in the South Pacific. Country Report no. 13: Palau*, Noumea: South Pacific Commission and International Labour Organisation.

—— (1983e) *Migration, Employment and Development in the South Pacific. Country Report no. 16: Solomon Islands*, Noumea: South Pacific Commission and International Labour Organisation.

—— (1983f) *Migration, Employment and Development in the South Pacific. Country Report no. 18: Tonga*, Noumea: South Pacific Commission and International Labour Organisation.

—— (1983g) *Migration, Employment and Development in the South Pacific. Country Report no. 19: Tuvalu*, Noumea: South Pacific Commission and International Labour Organisation.

—— (1983h) *Migration, Employment and Development in the South Pacific. Country Report no. 22: Western Samoa*, Noumea: South Pacific Commission and International Labour Organisation.

—— (1985a) *Migration, Employment and Development in the South Pacific. Country Report no. 4: Fiji*, Noumea: South Pacific Commission and International Labour Organisation.

—— (1985b) *Migration, Employment and Development in the South Pacific. Country Report no. 14: Papua New Guinea*, Noumea: South Pacific Commission and International Labour Organisation.

—— (1985c) *Migration, Employment and Development in the South Pacific. Country Report no. 20: Vanuatu*, Noumea: South Pacific Commission and International Labour Organisation.

—— (1986) 'Urban bias and development banking in the Kingdom of Tonga', *Pacific Economic Bulletin* 1(2): 22–6.

—— (1987a) 'Population growth and emigration: maintaining a balance', in M. Taylor (ed.) *Fiji: Future Imperfect*, Sydney: Allen and Unwin, 14–32.

—— (1987b) 'Migration, rural development and policy formation in the South Pacific', *Journal of Rural Studies* 3: 105–21.

—— (1988a) 'Temporary townsfolk? Siwai migrants in urban Papua New Guinea', *Pacific Studies* 11: 77–100.

—— (1988b) *Sovereignty and Survival: Island Microstates in the Third World*, Sydney: University of Sydney, Department of Geography, Monograph no. 3.

—— (1989) 'Migration, urbanisation et santé dans le Pacifique Sud', *Cahiers d'Outre-Mer* 42: 395–423.

—— (1990) 'Modernity and its discontents: migration and change in the South Pacific', in J. Connell (ed.) *Migration and Development in the South Pacific*, Canberra: Australian National University, National Centre for Development Studies, Pacific Research Monograph no. 24, 1–28.

—— (1993) *The Return Home? Albert Wendt, Migration and Identity in Polynesia*, Leeds: University of Leeds, School of Geography, Working Paper no. 93/23.

—— (1997a) *Papua New Guinea: The Struggle for Development*, London: Routledge.

—— (1997b) 'Health in Papua New Guinea: a decline in development', *Australian Geographical Studies* 35: 271–93.

—— (1999) 'Environmental change, economic development and emigration in Tuvalu', *Pacific Studies* 22: 1–20.

—— (2001) 'Island dreaming: the contemplation of Polynesian paradise', *Journal of Historical Geography* 27 (in press).

—— and Brown, R. (1995) 'Migration and remittances in the South Pacific: towards new perspectives', *Asian and Pacific Migration Journal* 4: 1–34.

—— and Curtain, R. (1982) 'Urbanization and inequality in Melanesia', in R.J. May and H. Nelson (eds) *Melanesia Beyond Diversity*, Canberra: Australian National University, 461–500.

—— and Lea, J.P. (1992) 'My country will not be there: global warming, development and the planning response in small island states', *Cities* 9: 295–309.

—— and —— (1993) *Planning the Future: Melanesian Cities in 2010*, Canberra: Australian National University, National Centre for Development Studies, Pacific Policy Paper no. 11.

—— and —— (1994) 'Cities of parts, cities apart? Changing places in modern Melanesia', *The Contemporary Pacific* 6: 267–309.

—— and —— (1995) *Urbanization in Polynesia*, Canberra: Australian National University, National Centre for Development Studies, Pacific Policy Paper no. 14.

—— and —— (1998) *Island Towns: Managing Urbanization in Micronesia*, Honolulu: Center for Pacific Islands Studies, University of Hawaii.

—— and Pritchard, W. (1990) 'Tax havens and global capitalism: Vanuatu and the Australian connection', *Australian Geographical Studies* 28: 38–50.

Conroy, J. (1982) *Essays on the Development Experience*, Port Moresby: Institute of Applied Social and Economic Research, Monograph no. 17.

Cornforth, R. (1992) 'Bridging the gap: bringing together customary and parliamentary law making in the management of natural resources in Western Samoa', unpublished paper to SPREP/UNEP Workshop, 'Strengthening Environmental Legislation in the Pacific Region', Apia.

Cotton, A. and Franceys, R. (1994) 'Infrastructure for the urban poor: policy and planning issues', *Cities* 11: 15–24.

Coyne, T. (1984) *The Effect of Urbanization and Western Diet on the Health of Pacific Island Populations*, Noumea: South Pacific Commission, Technical Paper no. 186.

Crocombe, R.G. (ed.) (1987) *Land Tenure in the Pacific*, 3rd edn, Suva: Institute of Pacific Studies.

—— (1995) *Customary Land Tenure and Sustainable Development: Complementarity or Conflict?*, Suva: Institute of Pacific Studies.

Curry, G. (1999) 'Markets, social embeddedness and precapitalist societies: the case of village tradestores in Papua New Guinea', *Geoforum* 30: 285–98.

Curtain, R. and Blaxter, L. (eds) (1978) *Papua New Guinea Public Housing Policy*, Port Moresby: Institute of Applied Social and Economic Research.

Dahanayake, P.A.S. (1991) *Housing Market Anomalies in Papua New Guinea*, Canberra: Australian National University, National Centre for Development Studies, Islands/Australia Working Paper no. 91/6.

Dening, G. (1980) *Islands and Beaches: Discourse on Silent Land. Marquesas 1774–1880*, Melbourne: Melbourne University Press.

Denoon, D. (1980) 'Closing the towns: solution or evasion?', in R.T. Jackson, P. Batho and J. Odongo (eds) *Urbanisation and its Problems in Papua New Guinea*, Port Moresby: University of Papua New Guinea Press, 283–9.

Dimitriou, H.T. (1991) 'An integrated approach to urban infrastructure development: a review of Indonesian experience', *Cities* 8: 192–208.

Dobbin, J. and Hezel, F.X. (1998) 'Sustainable human development in Micronesia', *Micronesian Counselor*, series 2, no. 4.

Dove, J., Miriung, T. and Togolo, M. (1974) 'Mining bitterness', in P. Sack (ed.) *Problems of Choice: Land in Papua New Guinea's Future*, Canberra: Australian National University Press, 181–5.

Drakakis-Smith, D.W. (1995) 'Third World cities: sustainable urban development 1', *Urban Studies* 32: 659–77.

Dyson, J. (1982) *The South Seas Dream*, London: Heinemann.

EDU International in association with GHK/MRM International (1992a) *Western Samoa Housing Sector Study: Final Report*, vol. 1: *Summary*, Apia: Government of Western Samoa and Asian Development Bank.

—— (1992b) *Western Samoa Housing Sector Study: Final Report*, vol. 2: *Sector Review and Recommendations*, Apia: Government of Western Samoa and Asian Development Bank.

—— (1992c) *Western Samoa Housing Sector Study: Final Report*, vol. 3: *Policies, Plans and Programmes*, Apia: Government of Western Samoa and Asian Development Bank.

Ehrlich, P. (1984) *Koror: A Center of Power, Commerce and Colonial Administration*, Saipan: Micronesian Archaelogical Survey Report no. 11.

Emberson-Bain, A. (1994a) 'Backbone of growth: export manufacturing and Fiji's tuna fish wives', in A. Emberson-Bain (ed.) *Sustainable Development or Malignant Growth: Perspectives of Pacific Island Women*, Suva: Marama Publications, 149–71.

—— (1994b) *Labour and Gold in Fiji*, Cambridge: Cambridge University Press.

—— (1995) *Women in Development: Kiribati*, Manila: Asian Development Bank.

EMPAT (1996) *Economic Use of Land in the Federated States of Micronesia: A Review and Description of Land Tenure Systems in the FSM*, Palikir: Economic Management Advisory Team.

Evans, J. (1992) *Transit of Venus*, London: Minerva.

Fahey, S. (1980) 'Urban commitment: a case study from Bulolo', in R.T. Jackson, P. Batho and J. Odongo (eds) *Urbanization and its Problems in Papua New Guinea*, Port Moresby: University of Papua New Guinea, 303–12.

Fairbairn, I. (1983) 'Second thoughts on first aid', *Islands Business* 9(11): 23–6.

—— (1992) *The Role of Small-scale Industry in Pacific Island Countries*, Sydney: University of New South Wales, Pacific Studies Monograph no. 4.

Fiji Council of Social Services (2000) *Citizens and Government: Civil Society in the New Millennium*, Suva.

Finau, S., Prior, I. and Maddill, J. (1987) 'Food consumption patterns among urban and rural Tongans', *Review* 15: 35–41.

Fischer, J. (1964) 'The abandonment of Nan Madol: ancient capital of Ponape', *Micronesica* 1: 49–54.

Flinn, J. (1990) 'We still have our customs: being Pulapese in Truk', in J. Linnekin and L. Poyer (eds) *Cultural Identity and Ethnicity in the Pacific*, Honolulu: University of Hawaii Press, 103–26.

—— (1997) 'Private houses, public sharing: Pollapese migrants coping with change', in J. Rensel and M. Rodman (eds) *Home in the Islands*, Honolulu: University of Hawaii Press, 132–50.

Flores, A. and Harris, G. (1982) 'The marketing of fresh fruit and vegetables in Port Moresby', *Pacific Viewpoint* 23: 147–60.

Forbes, D. (1975) 'Trends in vegetable marketing in Port Moresby', *South Pacific Bulletin* 25(1): 17–21.

—— and Jackson, R.T. (1975) 'Planning settlements: the example of Ranuguri, Port Moresby', *Australian Geographer* 13: 54–6.

Force, R. (1960) *Leadership and Cultural Change in Palau*, Chicago: Fieldiana Anthropology no. 50.

Forum Secretariat (1996) *Labour Market and Human Development Issues in the South Pacific Forum Region*, Suva: Forum Secretariat, Economic Development Division, Working Paper no. 96/1.

Fox, M. (1976) *Social Welfare Services and Development Planning in Pacific Islands*, Noumea: South Pacific Commission, Technical Paper no. 173.

Frazer, I. (1981) 'Man long taon: migration and differentiation among the To'ambaita, Solomon Islands', unpublished PhD thesis, Australian National University, Canberra.

—— (1997) 'Struggle for control of Solomon Islands forests', *The Contemporary Pacific* 9: 39–72.

Friedman, J. (1990) 'Being in the World: globalization and localization', *Theory, Culture and Society* 7: 311–28.

Fry, G. (1997) 'Framing the islands: knowledge and power in changing Australian images of the South Pacific', *The Contemporary Pacific* 9: 305–44.

Gaigo, B. (1970) 'Social disorder', in R.T. Jackson, P. Batho and J. Odongo (eds) *Urbanization and its Problems in Papua New Guinea*, Port Moresby: University of Papua New Guinea Press, 209.

Gailey, C.W. (1992) 'A good man is hard to find: overseas migration and the decentered family in the Tongan islands', *Critique of Anthropology* 12: 47–74.

Garnaut, R., Wright, M. and Curtain, R. (1977) *Employment, Incomes and Migration in Papua New Guinea Towns*, Port Moresby: Institute of Applied Social and Economic Research, Monograph no. 6.

Gewertz, D. (1978) 'Tit for tat: barter markets in the Middle Sepik', *Anthropological Quarterly* 51: 37–44.

—— and Errington, F. (1991) *Twisted Histories, Altered Contexts: Representing the Chambri in a World System*, Cambridge: Cambridge University Press.

—— and —— (1999) *Emerging Class in Papua New Guinea: The Telling of Difference*, Cambridge: Cambridge University Press.

Giddens, A. (1992) *The Consequences of Modernity*, Cambridge: Polity Press.

Gilbert, A. (1992) 'Third World cities: housing, infrastructure and servicing', *Urban Studies* 29: 435–60.

GKW Consultants (1989) *Tariff Study for the Water Supply Sector in Western Samoa: Final Report*, Apia: Public Works Department, Government of Western Samoa.

Goddard, M. (1992) 'Big man, thief: the social organization of gangs in Port Moresby', *Canberra Anthropology* 15: 20–34.

—— (1998) 'Off the record: village court praxis and the politics of settlement life in Port Moresby, Papua New Guinea', *Canberra Anthropology* 21: 41–62.

—— (2001) 'From rolling thunder to reggae: imagining squatter settlements in Papua New Guinea', *The Contemporary Pacific* 13: 1–32.

Gounis, C. and Rutz, H.J. (1986) 'Urban Fijians and the problem of urban unemployment', in C. Griffin and M. Monsell-Davis (eds) *Fijians in Town*, Suva: Institute of Pacific Studies, 50–88.

Government of Fiji and United Nations Development Programme (UNDP) (1997) *Fiji Poverty Report*, Suva: UNDP.

Government of Kiribati (1981) 'Urban drift and unemployment: problems and solutions', unpublished paper to 7th South Pacific Labour Ministers Conference, Tarawa.

Gregory, C. (1997) 'Hunting for water in Port Moresby', *Anthropology Today* 13 (4): 15–17.

Griffin, J., Nelson, H. and Frith, S. (1979) *Papua New Guinea: A Political History*, Melbourne: Heinemann.

Haberkorn, G. (1989) *Port Vila: Transit Station or Final Stop?*, Canberra: Australian National University, National Centre for Development Studies, Pacific Research Monograph no. 21.

Halliburton, T.W. (1992) *Urban Policy in Vanuatu: A Review*, Port Vila: United Nations Economic and Social Commission for Asia and the Pacific.

Hammar, L. (1992) 'Brothels, bamu and tu kina bus in south coast New Guinea', *Anthropology and Humanism* 21: 140–58.

Harré, J. and Knapman, C. (eds) (1977) *Living in Town: Problems and Priorities in Urban Planning in the South Pacific*, Suva: University of the South Pacific.

Harris, B. (1988) *The Rise of Rascalism: Action and Reaction in the Evolution of Rascal Gangs*, Port Moresby: Institute of Applied Social and Economic Research, Discussion Paper no. 24.

Harvey, D. (1989) *The Condition of Postmodernity*, Oxford: Basil Blackwell.

Haughton, G. (1997) 'Developing sustainable urban development models', *Cities* 14: 189–95.

—— and Hunter, C. (1994) *Sustainable Cities*, London: Jessica Kingsley.

Hau'ofa, E. (1977) *Our Crowded Islands*, Suva: Institute of Pacific Studies, University of the South Pacific.

Hezel, F.X. (1983) *The First Taint of Civilization: A History of the Caroline and Marshall Islands in Pre-colonial Days, 1521–1885*, Honolulu: University of Hawaii Press.

—— (1985) 'Sale of land', *Justice Issues in Micronesia*, no. 1, Pohnpei, FSM.

—— (1994) 'Land issues in Chuuk', *Micronesian Counselor*, no. 16, Pohnpei, FSM.

—— (1995) *Strangers in Their Own Land: A Century of Colonial Rule in the Caroline and Marshall Islands*, Honolulu: University of Hawaii Press.

—— (1999) 'Land: is it time for a change in direction?', *Micronesian Counselor*, no. 23, Pohnpei, FSM.

—— and Levin, M. (1996) 'New trends in Micronesian migration: FSM migration to Guam and the Marianas, 1990–1993', *Pacific Studies* 19: 91–114.

Hildebrand, S.F. (1999) 'Sewage – a problem to not be treated lightly', *Pacific Islands Monthly* 69(1): 24–5.

Hobsbawm, E. and Ranger, T. (1983) *The Invention of Tradition*, Cambridge: Cambridge University Press.

Home, R. (1993) 'Transferring British planning law to the colonies', *Third World Planning Review* 15: 397–410.

Honiara Town Council (1988) *Honiara Town Council Development Plan 1988–1992*, Honiara.

Howard, D. (1976) *Problems of Social Deviance in Tuvalu*, London: Overseas Development Administration.

Howlett, D. (1973) *Papua New Guinea: Geography and Change*, Melbourne: Nelson.

——, Hide, R. and Young, E. (1976) *Chimbu: Issues in Development*, Canberra: Australian National University, National Centre for Development Studies, Monograph no. 4.

Howorth, R. (2000) 'Because of the sickness, "no sashimi"', *Islands Business*, 26(8): 51.

Hughes, I. (1977) *New Guinea Stone Age Trade*, Canberra: Terra Australis no. 3.

Hunt, C. (1998) *Pacific Development Sustained: Policy for Pacific Environments*, Canberra: Australian National University, National Centre for Development Studies, Pacific Policy Paper no. 32.

Inglis, A. (1974) *Not a White Woman Safe: Sexual Anxiety and Politics in Port Moresby 1920–1934*, Canberra: Australian National University Press.

Jabre, B., Raoult, A., Pritchard, C. and Speake, J. (1976) *Study of the Suburban District of Tagabe (Vila, New Hebrides)*, Noumea: South Pacific Commission.

Jackson, R.T. (1976a) 'Lae', in R.T. Jackson (ed.) *An Introduction to the Urban Geography of Papua New Guinea*, Port Moresby: University of Papua New Guinea, Department of Geography, 274–304.

—— (1976b) 'Bulolo and Wau', in R.T. Jackson (ed.) *An Introduction to the Urban Geography of Papua New Guinea*, Port Moresby: University of Papua New Guinea, Department of Geography, 305–21.

—— (1976c) 'Rabaul', in R.T. Jackson (ed.) *An Introduction to the Urban Geography of Papua New Guinea*, Port Moresby: University of Papua New Guinea, Department of Geography, 415–37.

—— (ed.) (1976d) *An Introduction to the Urban Geography of Papua New Guinea*, Port Moresby: University of Papua New Guinea, Department of Geography, Occasional Paper no. 13.

—— (1978) 'Housing trends and policy implications in Papua New Guinea: flaunting the flag of abstracted empiricism', in P. Rimmer, D. Drakakis-Smith and T. McGee (eds) *Food, Shelter and Transport in Southeast Asia and the Pacific*, Canberra: Australian National University, Department of Human Geography, 171–87.

—— (1980) 'What's wrong with towns', in R.T. Jackson, P. Batho and J. Odongo (eds) *Urbanisation and its Problems in Papua New Guinea*, Port Moresby: University of Papua New Guinea, 111–17.

James, K.E. (1991) 'Migration and remittances: a Tongan village perspective', *Pacific Viewpoint* 32: 1–23.

—— (1993) 'Cutting the ground from under them? Commercialization, cultivation and conservation in Tonga', *The Contemporary Pacific* 5: 215–42.

—— (1995) 'Right and privilege in Tongan land tenure', in R.G. Ward and E. Kingdon (eds) *Land, Custom and Practice in the South Pacific*, Cambridge: Cambridge University Press, 157–97.

Jayaraman, T.K. (1995) *Demographic and Socioeconomic Determinants of Contraceptive Use Among Urban Women in the Melanesian Countries in the South Pacific: A Case Study of Port Vila in Vanuatu*, Manila: Asian Development Bank.

Johannes, R.E. (1981) *Words of the Lagoon: Fishing and Marine Lore in the Palau District of Micronesia*, Berkeley and Los Angeles: University of California Press.

Johnson, R. and Threlfall, N. (1985) *Volcano Town: The 1937–43 Eruptions at Rabaul*, Bathurst: Robert Brown.

Johnston, R. (1999) 'Water crisis', *Islands Business* 25 (2): 24–7.

Jones, P. (1994) *Urban Management in South Tarawa Republic of Kiribati*, Bairiki: Ministry of Home Affairs and Rural Development.

—— (1995) *Responding to Urban Change: Observations on Urbanization, Urban Planning and Urban Management in the Pacific*, Bairiki: Government of Kiribati Printer.

—— (1996) 'Institutional strengthening of the Land and Survey Division, Ministry of Home Affairs and Rural Development, Republic of Kiribati', unpublished report, Canberra: AusAID.

Jourdan, C. (1994) 'Créolisation, urbanisation et identité aux îles Salomon', *Journal de la Société des Océanistes* 99: 177–86.

Kama, T. (1979) 'Guadalcanal Plains', in J. Heath (ed.) *Land Research in Solomon Islands*, Honiara: Ministry of Agriculture and Lands, 150–9.

Kane, P. and Lucas, D. (1986) *Population Problems in the South Pacific Islands: An Overview*, Canberra: Australian National University, National Centre for Development Studies.

Keesing, R. (1982) '*Kastom* in Melanesia: an overview', *Mankind* 13: 297–301.

—— (1994) 'Foraging in the urban jungle: notes from the Kwaio underground', *Journal de la Société des Océanistes* 99: 167–75.

King, A. (1990) *Urbanism, Colonialism and the World Economy*, London: Routledge.

King, D. (1983) *The Small Towns and Outstations of Papua New Guinea*, Port Moresby: University of Papua New Guinea, Department of Geography, Occasional Paper no. 3.

—— (1992) 'Socio-economic differences between residential categories in Port Moresby', *Yagl-Ambu* 16 (3): 1–14.

Kinhill Riedel and Byrne in association with PPK Consultants (1993) *Management Plans for the Fanga'uta Lagoon System and/or Low-lying Areas of the Kingdom of Tonga*, Canberra: Australian International Development Assistance Bureau.

Kiribati Housing Corporation (1996) *The State of Shelter in South Tarawa*, Tarawa: Kiribati Housing Corporation.

—— (1997) *Draft Kiribati National Shelter Strategy*, Tarawa: Kiribati Housing Corporation.

Kiribati Ministry of Home Affairs and Rural Development (KMHARD) (1993) *Challenging the Future: Towards Development Standards for Land, Housing, Water, Sewage and Waste in South Tarawa*, Bairiki: KMHARD.

—— (1995) *Urban Management Plan for South Tarawa, Republic of Kiribati*, Bairiki: Government of Kiribati Printer.

—— (2000a) *Kiribati Land Information System: A Manual Providing an Overview of KLIS*, South Tarawa: KMHARD.

—— (2000b) *Urban Planning Policy and Procedures Manual*, South Tarawa: KMHARD.

Klinckhamers, P. (1992) *Western Samoa's Land Based Pollution Sources and Their Effects on the Marine Environment*, Apia: South Pacific Regional Environmental Programme.

Knapman, B. and Saldanha, C. (1999) *Reforms in the Pacific*, Manila: Asian Development Bank.

Koch, K.-F. (1978) *Logs in the Current of the Sea*, Canberra: Australian National University Press.

Koczberski, G., Curry, G. and Connell, J. (2001) 'Full circle or spiralling out of control? State violence and the control of urbanisation in Papua New Guinea', *Urban Studies* 38: 2017–36.

Lal, B. (2000) '"Chiefs and thieves and other people besides": the making of George Speight's coup', *Journal of Pacific History* 35: 281–93.

Larmour, P. (1982) *Report on Urban Management in South Tarawa, Kiribati*, Suva: UN Development Advisory Team.

—— (1995) 'Governance, governmentality and South Pacific aid', in *Governance and Good Government: Policy and Implementation in the South Pacific*, Canberra: Australian National University, National Centre for Development Studies, 17–20.

—— (ed.) (1997) *The Governance of Common Property in the Pacific Region*, Canberra: Australian National University, National Centre for Development Studies.

Lateef, S. (1990) 'Current and future implications of the coups for women in Fiji', *Contemporary Pacific* 2: 113–30.

Lawrence, R. (1992) *Tamana Fifteen Years On: A Survey of Changes in the Household Economy since 1974*, Wellington: Department of Geography, Victoria University.

Lea, D.A.M. and Gray, N. (1982) 'Enga demography', in B. Carrad, D. Lea and K. Talyaga (eds) *Enga: Foundations for Development*, Armidale, NSW: University of New England, 41–58.

Lea, J.P. (1979) 'Self-help and autonomy in housing: theoretical critics and empirical investigators', in H. Murison and J.P. Lea (eds) *Housing in Third World Countries: Perspectives on Policy and Practice*, London: Macmillan, 49–53.

—— (1983) 'Customary tenure and urban housing land: partnership and participation in developing societies', in S. Angel *et al.* (eds) *Land for Housing the Poor*, Singapore: Select Books, 54–72.

Leaney, C. (1966) 'Housing survey of New Hebridean natives and other Pacific islanders resident in the Vila area', in *Second Regional Conference on Housing in the South Pacific*, Noumea: South Pacific Commission, Technical Paper.

Lee, Y.F. (1997) 'The privatization of solid waste infrastructure and services in Asia', *Third World Planning Review* 19: 139–62.

Leupena, T. and Lutelu, K. (1987) 'Providing for the multitude', in R. Crocombe (ed.) *Land Tenure in the Atolls*, Suva: University of the South Pacific, 143–65.

Levantis, T. (1997a) 'The urban labour market in Papua New Guinea, post deregulation', *Pacific Economic Bulletin* 12 (2): 54–72.

—— (1997b) 'Urban unemployment in Papua New Guinea: it's criminal', *Pacific Economic Bulletin* 12 (2): 73–84.

—— (2000) 'Crime catastrophe: reviewing Papua New Guinea's most serious social and economic problem', *Pacific Economic Bulletin* 15 (1): 130–42.

Levine, H.B. and Levine, M.W. (1979) *Urbanization in Papua New Guinea: A Study of Ambivalent Townsmen*, Cambridge: Cambridge University Press.

Lewis, G. (1980) *Days of Shining Red*, Cambridge: Cambridge University Press.

Lieber, M. (1977) 'The processes of change in two Kapingamarangi communities', in *Exiles and Migrants in Oceania*, Hawaii: University of Hawaii Press, 35–67.

Love, R.S. (1982) 'Basic needs in Kiribati', *Pacific Perspective* 10(2): 66–74.

Luana, C. (1969) 'Buka: a retrospect', *New Guinea* 4 (1), 15–20.

Ludwig, H.F. and Browder, G. (1992) 'Appropriate water supply and sanitation technology for developing countries in tropical monsoon climates', *Environmentalist* 12: 131–9.

McArthur, M. (1981) *New Hebrides Population 1840–1967: A Re-interpretation*, Noumea: South Pacific Commission, Occasional Paper no. 18.

McDaniel, C. and Gowdy, J. (2000) *Paradise for Sale: A Parable of Nature*, Berkeley: University of California Press.

Macdonald, B. (1982) *Cinderellas of the Empire: Towards a History of Kiribati and Tuvalu*, Canberra: Australian National University Press.

McGavin, P.A. (1991) *Wages, Incomes and Productivity in Papua New Guinea*, Port Moresby: Institute of National Affairs, Discussion Paper no. 48.

McGee, T.G., Ward, R.G. and Drakakis-Smith, D.W. (1980) *Food Distribution in the New Hebrides*, Canberra: Australian National University, National Centre for Development Studies, Monograph no. 25.

McGrath, W. (1970) 'The effects of urban drift in the Palau district of Micronesia', in M.W. Ward (ed.) *Change and Development in Rural Melanesia*, Canberra: Australian National University Press, 10–33.

McIlwaine, C. (1997) 'Third World development: urbanizing for the future', *Progress in Human Geography* 21: 406–13.

McInnes, L. and Connell, J. (1988) 'The world system in a Fijian store', *South Pacific Forum* 4: 116–21.

McMurray, C. (1992) 'Issues in population planning: the case of Papua New Guinea', *Development Bulletin* 24: 13–16.

Macpherson, C. (1992) 'Economic and political restructuring and the sustainability of migrant remittances: the case of Western Samoa', *Contemporary Pacific* 4: 109–35.

—— and Macpherson, L. (1987) 'Toward an explanation of recent trends in suicide in Western Samoa', *Man* 22: 305–30.

Mamak, A. (1978) *Colour, Culture and Conflict: A Study of Pluralism in Fiji*, Sydney: Pergamon Press.

Marcus, G.E. (1981) 'Power on the extreme periphery: the perspective of Tongan elites in the modern world system', *Pacific Viewpoint* 22: 48–64.

Marshall, M. (1979a) *Weekend Warriors: Alcohol in a Micronesian Culture*, Palo Alto, CA: Mayfield.

—— (1979b) 'Education and depopulation on a Micronesian atoll', *Micronesica* 15: 1–11.

Matheson, T. (1987) 'Niuean land-holdings in Western Samoa', in L. Mason and P. Hereniko (eds) *In Search of a Home*, Suva: Institute of Pacific Studies, 65–83.

Maude, A. and Sevele, F. (1987) 'Tonga: equality overtaking privilege', in R. Crocombe (ed.) *Land Tenure in the Pacific*, Suva: University of the South Pacific, 114–42.

Maude, H.E. (1981) *Slavers in Paradise: The Peruvian Labour Trade in Polynesia, 1862–1864*, Suva: Institute of Pacific Studies.

Mecartney, S. (2001) 'Blacksands settlement: a case for urban permanence in Vanuatu', unpublished MSc thesis, University of Sydney.

Meleisea, M. (1992) *Change and Adaptations in Western Samoa*, Christchurch: University of Canterbury, Macmillan Brown Centre for Pacific Studies.

Metcalf, J.P. (1969) *Manpower Assessment in the New Hebrides*, Geneva: International Labour Organisation.

Miles, R., Alam, M., Boyle, J. and Larhed, T. (1992) *Employment in Western Samoa: Present and Potential*, Suva: International Labour Organisation.

Miskaram, N. (1976) 'Kavieng', in R.T. Jackson (ed.) *An Introduction to the Urban Geography of Papua New Guinea*, Port Moresby: University of Papua New Guinea, Department of Geography, Occasional Paper no. 13, 397–414.

Mitchell, J. (2000) 'Violence as continuity: violence as rupture. Narratives from an urban settlement in Vanuatu', in S. Dinnen and A. Ley (eds) *Reflections on Violence in Melanesia*, Sydney: Hawkins Press, 189–208.

Mitchell, T. (1988) *Colonizing Egypt*, Cambridge: Cambridge University Press.

Mitlin, D. and Satterthwaite, D. (1996) 'Sustainable development and cities', in C. Pugh (ed.) *Sustainability, the Environment and Urbanization*, London: Earthscan, 23–61.

Monsell-Davis, M. (1993) 'Urban exchange: safety net or disincentive? *Wantoks* and relatives in the Urban Pacific', *Canberra Anthropology* 16: 45–66.

—— (2000) 'Social change, contradictions, youth and violence in Fiji', in S. Dinnen and A. Ley (eds) *Reflections on Violence in Melanesia*, Sydney: Hawkins Press, 209–22.

Moorhead, A. (1966) *The Fatal Impact: The Invasion of the South Pacific 1767–1840*, London: Hamish Hamilton.

Morauta, L. (1980) 'Permanent urban residents in Papua New Guinea: problems and prospects', in R.T. Jackson, P. Batho and J. Odongo (eds) *Urbanization and its Problems in Papua New Guinea*, Port Moresby: University of Papua New Guinea, 290–302.

—— and Ryan, D. (1982) 'From temporary to permanent townsmen: migrants from the Malalaua district, Papua New Guinea', *Oceania* 53: 39–55.

——, Igo, J., Larmour, P. and Regan, A. (1986) *Government for the NCD. Phase I: Final Report*, Port Moresby.

Morrison, R.J. and Munro, A.J. (1999) 'Waste management in the small island developing states of the South Pacific: an overview', *Australian Journal of Environmental Management* 6: 232–46.

Mount Hagen Town Authority (1991) *Annual Report 1990*, Mount Hagen: Mount Hagen Town Authority.

Munro, D. and Bedford, R. (1980) 'Historical backgrounds', in *A Report on the Results of the Census of Tuvalu*, Funafuti: Government of Tuvalu, 1–13.

Nage, J. (1987) 'Immigrant settlements in Honiara, Solomon Islands', in L. Mason and P. Hereniko (eds) *In Search of a Home*, Suva: Institute of Pacific Studies, 93–102.

Naibavu, T. and Schutz, B. (1986) 'Prostitution: problem or profitable industry?', in C. Griffin and M. Monsell-Davis (eds) *Fijians in Town*, Suva: Institute of Pacific Studies, 89–101.

Naidu, V. (1981) 'Fijian development and national unity: some thoughts', *Review* 2: 3–12.

Nair, S. (1980) *Rural-born Fijians and Indo-Fijians in Suva: A Study of Movements and Linkages*, Canberra: Australian National University, National Centre for Development Studies, Monograph no. 24.

Nardi, B. (1983) 'Goals in reproductive decision-making', *American Ethnologist* 10: 697–714.

Nathan Associates (1966) *Economic Plan for Micronesia*, Washington, DC: Nathan Associates.

Nayacakalou, R. (1963) 'The urban Fijians of Suva', in R. Spoehr (ed.) *Pacific Port Towns and Cities*, Honolulu: Bishop Museum Press, 33–41.

—— (1971) 'Fiji: manipulating the system', in R. Crocombe (ed.) *Land Tenure in the Pacific*, London: Oxford University Press, 206–26.

Nelson, H. (1976) *Black, White and Gold: Goldmining in Papua New Guinea*, Canberra: Australian National University Press.

Nevin, D. (1977) *The American Touch in Micronesia*, New York: W.W. Norton.

Nibbrig, N. (1992) 'Rascals in paradise: urban gangs in Papua New Guinea', *Pacific Studies* 15: 115–34.

Norwood, H. (1979) 'Port Moresby: patterns of settlement among migrant and urban villagers', in C.A. and B. Valentine (eds) *Going Through Changes: Villagers, Settlers and Development in Papua New Guinea*, Port Moresby: Institute of Papua New Guinea Studies, 73–89.

—— (1983) *Port Moresby: Urban Villages and Squatter Areas*, Port Moresby: University of Papua New Guinea, Department of Geography.

O'Carroll, J. (1997) 'Multiple cities: Suva and the (post) colonial', in S. Mishra and E. Guy (eds) *Dreadlocks in Oceania*, Suva: Department of Languages and Literature, University of the South Pacific, 26–54.

Oeser, L. (1969) *Hohola: The Significance of Social Networks in Urban Adaptation of Women in Papua New Guinea's First Low-cost Housing Estate*, Canberra: New Guinea Research Unit, Australian National University.

Ogden, M.R. (1994) 'MIRAB and the Marshall Islands', *Isla* 2: 237–72.

Oliver, D.L. (1951) *Planning Micronesia's Future*, Cambridge, MA: Harvard University Press.

Olsen, R.Y. (1991) *Infrastructure Services in Fiji*, Washington, DC: Planning and Development Collaborative International.

O'Meara, J.T. (1995) 'From corporate to individual land tenure in Western Samoa', in R.G. Ward and E. Kingdon (eds) *Land, Custom and Practice in the South Pacific*, Cambridge: Cambridge University Press, 109–56.

Ono, K. (2002) 'Building paradise: the establishment of Japanese colonial towns in the Western Pacific', unpublished PhD thesis, University of Sydney.

Oram, N. (1976) *Colonial Town to Melanesian City*, Canberra: Australian National University Press.

—— (1980) 'Land, housing and administration in Honiara: towards a concerted policy', *O'o: A Journal of Solomon Island Studies* 1: 133–63.

Osborne, D. (1966) 'The archaeology of the Palau Islands: an intensive study', Honolulu: *Bishop Museum Bulletin* 230.

Overton, J. (1992) 'The limits to accumulation: changing land tenure in Fiji', *Journal of Peasant Studies* 19: 326–42.

Owen, R. (1955) *Where the Poor are Happy*, London: Collins.

Papua New Guinea. Bureau of Statistics (1980) *Urban Population Survey 1977: Housing, Population, Education, Training, Economic Activity and Migration*, Port Moresby: Bureau of Statistics, Bulletin No. 8.

Papua New Guinea. Department of Lands and Physical Planning (1988) *Land Mobilisation Programme*, Port Moresby: Government Printer.

Papua New Guinea. National Housing Corporation (1992) *Official National Housing Corporation Handbook*, Port Moresby: National Housing Corporation.

Parkinson, S.V. (1977) 'Some observations on the cause of malnutrition in Pacific island urban communities', in J. Harré and C. Knapman (eds) *Living in Town*, Suva: University of the South Pacific, 69–75.

Parliament of Fiji (1979) *Report on the Census of the Population*, Suva: Parliamentary Paper no. 43.

Parry, G.L. (1972) 'Organized juvenile crime in Port Moresby', *South Pacific Bulletin* 22 (1): 43–4, 60.

Peattie, M. (1988) *Nan'yo: The Rise and Fall of the Japanese in Micronesia, 1885–1945*, Honolulu: University of Hawaii Press.

Pintz, W. (1984) *Ok Tedi: Evolution of a Third World Mining Project*, London: Mining Journal Books.

Pirie, P. (1994) *Demographic Transition in the Pacific Islands: The Situation in the Early*

1990s, Honolulu: East-West Center, Pacific Islands Development Series, Working Paper no. 5.

Pohnpei State Land Use and Zoning Master Plan Taskforce (1995) *Pohnpei State Land Use and Zoning Master Plan Working Draft*, Kolonia: Department of Land.

Poirine, B. (1998) 'Should we hate or love MIRAB?', *The Contemporary Pacific* 10: 65–105.

Polier, N. (1966) 'Of mines and min: modernity and its malcontents in Papua New Guinea', *Ethnology* 35: 1–16.

Pollock, N. (1975) 'The risks of dietary change: a Pacific atoll example', in R.W. Casteel and G. Quimby (eds) *Maritime Adaptations of the Pacific*, The Hague: Mouton, 255–64.

—— (1996) 'Namu atoll revisited: a follow-up study of 25 years of resource use', *Atoll Research Bulletin* 440: 1–12.

Po'o, T. (1975) 'Gangs in Port Moresby', *Administration and Development* 3: 30–7.

Power, J., Wettenhall, R. and Halligan, J. (1981) *Local Government Systems of Australia*, Canberra: Australian Government Publishing Service.

Poyer, L. (1993) *The Ngatik Massacre: History and Identity on a Micronesian Atoll*, Washington, DC: Smithsonian Institute Press.

PPK Consultants in association with Riedel and Byrne (1991) *Neiafu Water Supply Draft Master Plan*, Canberra: Australian International Development Assistance Board.

—— (1992a) *Tonga Water Supply Master Plan Project: Institutional and Community Component Master Plan*, Canberra: Australian International Development Assistance Board.

—— (1992b) *Tonga Water Supply Master Plan Project: Nuku'alofa Master Plan*, Canberra: Australian International Development Assistance Board.

—— (1992c) *Tonga Water Supply Master Plan Project: Executive Summary*, Canberra: Australian International Development Assistance Board.

Price, W. (1966) *America's Paradise Lost*, New York: John Day.

Pringle, G. (1989) 'Heritage assessment: Apia, Western Samoa', unpublished MSc thesis, Department of Architecture, University of Sydney.

Pulea, M. (1992) *Kingdom of Tonga: Review of Environmental Law*, Apia: South Pacific Regional Environmental Programme.

Ralston, C. (1977) *Grass Huts and Warehouses: Pacific Beach Communities of the Nineteenth Century*, Canberra: Australian National University Press.

Ratcliffe, J. and Dillon, R. (1982) *A Review and Study of the Human Settlements Situation in the Kingdom of Tonga*, Nuku'alofa: Central Planning Department.

Rawlings, G. (1999a) 'Foundations of urbanisation: Port Vila Town and Pango Village, Vanuatu', *Oceania* 70: 72–88.

—— (1999b) 'Villages, islands and tax havens: the global/local implications of a financial entrepot in Vanuatu', *Canberra Anthropology* 22 (2): 37–50.

Ray, B. (1998) *Good Governance, Administrative Reform and Socioeconomic Realities: A South Pacific Perspective*, Canberra: Australian National University, State, Society and Governance in Melanesia Discussion Paper no. 98/2.

Reddy, K.M. (1976) *Some Squatters in Suva*, Suva: University of the South Pacific, School of Social and Economic Development, Basic Research in Pacific Islands Geography no. 4.

Rees, W.E. (1992) 'Ecological footprints and appropriated carrying capacity', *Environment and Urbanization* 4: 121–30.

Rehuher, F. (1993) 'Comments on youth development and movement: providing an alternative to urban drift, a Palau case', in *The Family in the Aquatic Continent*, Maui: Maui Pacific Center, 21–4.

Robert, D. (1993) *Water Authority Organization and Tariff Study, Draft Report*, Apia: Government of Western Samoa and Commission of the European Community.

Rodman, M. (1995) 'Breathing spaces: customary land tenure in Vanuatu', in R.G. Ward and E. Kingdon (eds) *Land, Custom and Practice in the South Pacific*, Cambridge: Cambridge University Press, 65–108.

Roth, G. (1987) *The Private Provision of Public Services in Developing Countries*, New York: World Bank and Oxford University Press.

Ryan, D. (1989) 'Home ties in town: Toaripi in Port Moresby', *Canberra Anthropology* 12: 19–27.

SAGRIC International (1996) *Palau National Master Development Plan, Revised Draft Final Report*, Adelaide: SAGRIC International and the Republic of Palau.

Samasoni, S. (1987) 'Land rights of the Niuean community in Apia', in L. Mason and P. Hereniko (eds) *In Search of a Home*, Suva: University of the South Pacific, 84–90.

Samy, J. (1977) 'Squatters, insecurity and housing development', in J. Harré and C. Knapman (eds) *Living in Town*, Suva: University of the South Pacific, 50–6.

Sato, M. (1997) 'Structure and dynamics of MIRAB societies in the South Pacific: an economic study', in K. Sudo and S. Yashida (eds) *Contemporary Migration in Oceania: Diaspora and Network*, Osaka: Japan Centre for Area Studies, 165–78.

Saw, C. and Belz, L. (1997) *Controlled Solid Waste Landfill in Nuku'alofa, Tonga*, Manila: WHO, Healthy Cities – Healthy Islands Document no. 8.

Schoeffel, P. (1996) *Sociocultural Issues and Economic Development in the Pacific Islands*, Manila: Asian Development Bank.

—— (1997) *Myths of Community Management: Sustainability, the State and Rural Development in Papua New Guinea, Solomon Islands and Vanuatu*, Canberra: Australian National University, State, Society and Governance in Melanesia Discussion Paper no. 97/8.

Secretariat of the Pacific Community (2001) *Pacific Island Populations*, Noumea: South Pacific Commission.

Seers, D. (1983) *The Political Economy of Nationalism*, Oxford: Oxford University Press.

Sevele, F. (1979) *Rural Employment and Rural Development in the Pacific Islands*, Noumea: South Pacific Commission.

Shankman, P. (1976) *Migration and Underdevelopment: The Case of Western Samoa*, Boulder, CO: Westview Press.

—— (1993) 'The Samoan exodus', in V. Lockwood, T. Harding and B. Wallace (eds) *Contemporary Pacific Societies: Studies in Development and Change*, Englewood Cliffs, NJ: Prentice-Hall, 156–70.

Shore, B. (1978) 'Introduction', in C. Macpherson, B. Shore and R. Franco (eds) *New Neighbors: Islanders in Adaptation*, Santa Cruz, CA: Center for South Pacific Studies, xiii–xiv.

Slater, D. (1986) 'Capitalism and urbanisation at the periphery: problems of interpretation and analysis with reference to Latin America', in D.W. Drakakis-Smith (ed.) *Urbanisation in the Developing World*, London: Routledge, 7–21.

—— (1992) 'Theories of development and politics of the postmodern: exploring a border zone', *Development and Change* 23: 283–319.

Slatter, C. (1987) 'Women factory workers in Fiji: the "half a loaf" syndrome', *Journal of Pacific Studies* 13: 47–59.

Smith, D.R. (1996) 'Palauans in Guam', *Micronesian Counselor*, August: 6–20.

Smith, M.P. (1992) 'Postmodernism, urban ethnography and the new social space of ethnic identity', *Theory and Society* 21: 493–531.

Sofield, T.H.B. (1993) 'Indigenous tourism development', *Annals of Tourism Research* 20: 729–50.

Somare, M. (1973) 'New goals for Papua New Guinea', *Pacific Perspectives* 2: 1–4.

Soulsby, J. and Murray, W. (2000) 'Urban agriculture, food security and the new urban poor', unpublished mimeo, Suva.

South Pacific Commission (1992) *The Pacific Way*, Noumea: South Pacific Commission.

—— (1994) *Pacific Island Populations*, Noumea: South Pacific Commission.

—— (1997) *Pacific Island Populations Data Sheet*, Noumea: South Pacific Commission.

Standing, G. (1982) *Circulation and Proletarianisation*, Geneva: International Labour Organisation, WEP Working Paper no. 119.

Stavenuiter, S. (1983) *Income Distribution in Fiji*, Suva: Central Planning Office and International Labour Organisation.

Storch, G., Gunn, R., Martin, W., Pollard, R. and Sinclair, S. (1980) 'Shigellosis in the Marshall Islands: epidemiological aspects of an outbreak', *American Journal of Tropical Medicine* 29: 456–63.

Storey, D. (1997) 'The political economy of urban development in the Pacific century: urbanization, policy responses and emerging discord in Polynesian towns', unpublished paper to VIII Pacific Science Inter-Congress, University of the South Pacific, Suva.

—— (1998a) 'Managing urban decline? Urban governance in the context of Port Vila, Vanuatu', *Development Bulletin* 45: 32–5.

—— (1998b) 'The politics of managing urban development in Pacific island states: the case of Samoa and Tonga', *Journal of Pacific Studies* 22, 61–80.

—— (1999) 'Sustainability and the urban Pacific: the case of Samoa and Tonga', in J. Overton and R. Scheyvens (eds) *Strategies for Sustainable Development*, Sydney: University of New South Wales Press, 155–67.

Strathern, M. (1975) *No Money on Our Skins: Hagen Migrants in Port Moresby*, Port Moresby and Canberra: New Guinea Research Bulletin no. 61.

—— (1977) 'The disconcerting tie: attitude of Hagen migrants towards home', in R.J. May (ed.) *Change and Movement*, Canberra: Australian National University Press, 247–66.

Stretton, A. (1979) *Urban Housing Policy in Papua New Guinea*, Port Moresby: Institute of Applied Social and Economic Research.

Sutherland, W. (2000) 'Global imperatives and economic reform in the Pacific island states', *Development and Change* 31: 459–80.

Takaio, A. (1993) 'Kain Tarawa', in H. van Trease (ed.) *Atoll Politics: The Republic of Kiribati*, Suva: Institute of Pacific Studies, 278–84.

Talyaga, K. and Olela, C. (1978) *The Debate on Repatriation of Squatters and Migrants from the North Solomon Province*, Port Moresby: Institute of Applied Social and Economic Research.

Tara, T. (1996) 'Prostitution: a growing problem in Honiara', *Islands Business* 22 (8): 22–3.

Thistlethwaite, R. and Davis, D. (1996) *A Sustainable Future for Melanesia? Natural Resources, Population and Development*, Canberra: Australian National University, National Centre for Development Studies, Pacific Policy Paper no. 17.

Thomas, P. (1984) 'Western Samoa: custom, change and constraints in relation to land rights, registration and productivity', in B. Acquaye and R. Crocombe (eds) *Land Tenure and Rural Productivity in the Pacific Islands*, Suva: University of the South Pacific, 135–54.

Tiffany, S.W. (1979) 'Port town village organisation in Western Samoa', *Journal of the Polynesian Society* 88: 127–75.

Toft, S. (ed.) (1997) *Compensation for Resource Development in Papua New Guinea*, Canberra: Australian National University, National Centre for Development Studies, Pacific Policy Paper no. 24.

To Mutnaram, M. (1979) 'Rescuing Papua New Guinea's urban housing crisis', unpublished paper to Waigani Seminar, Port Moresby.

Tonga Central Planning Department (1991) *Kingdom of Tonga Sixth Development Plan: 1991–1995*, Nuku'alofa: Central Planning Department.

Topping, D. (1980) 'Marshalls: eager to get it over', *Pacific Islands Monthly* 52 (1): 13–15.

UN Centre for Human Settlements (UNCHS) (1996) *An Urbanizing World: Global Report on Human Settlements*, Oxford: Oxford University Press.

—— and UN Development Programme (UNDP) (1994) *Review of Human Settlements in Pacific Atoll Nations: Republic of the Marshall Islands, Republic of Kiribati and Tuvalu*, Nairobi: UNCHS and UNDP.

UN Development Programme (UNDP) (1993) *Report of Joint UNDP/SPC Project Formulation Mission to Western Samoa*, Suva: UNDP.

—— (1994) *Pacific Human Development Report*, Suva: UNDP.

—— (1997) *Sustaining Livelihoods. Promoting informal sector growth in Pacific island countries*, Suva: UNDP.

—— (1999) *Pacific Human Development Report*, Suva: UNDP.

—— and UN Centre for Human Settlements (UNCHS) (1991) *Policy Recommendations for Housing and Urban Development*, Suva: Ministry of Housing and Urban Development.

Va'a, L. (1992) 'The future of Western Samoan migration to New Zealand', *Asia Pacific Migration Journal* 1: 313–32.

Vanderburg, P. (1992) 'Apia: town without a plan', *O le Siosiomaga Society Newsletter*, April: 7–8.

Vavine, P.W. (1984) 'Ivane settlement: a case study of first generation migrants to Port Moresby, Papua New Guinea', *Ambio* 13: 309.

Vitarelli, W.V. (1967) *Operation Exodus: A Program for Controlling the Population of Ebeye*, Saipan: Trust Territory of the Pacific Islands.

Walsh, A.C. (1964) 'Nuku'alofa, Tonga: a preliminary study of urbanisation and immigration', unpublished MA thesis, Victoria University, Wellington.

—— (1979) 'Appropriate and inappropriate housing for the urban poor: the Fiji experience', unpublished paper to ANZAAS Conference, Auckland.

—— (1982a) *Migration, Urbanization and Development in South Pacific Countries*, Bangkok: UN Economic and Social Commission for Asia and the Pacific.

—— (1982b) *Street Vending in Port Moresby, 1982*, Port Moresby: University of Papua New Guinea, Department of Geography.

—— (1984) 'The search for an appropriate housing policy in Fiji', *Third World Planning Review* 6: 185–200.

Ward, A. (1994) 'Managing indigenous land in the South Pacific', *Development Bulletin* 31: 47–9.

Ward, Marion (1970) 'Urbanisation: threat or promise?', *New Guinea* 5 (1): 57–62.

—— (1995) *Women and Employment in Solomon Islands*, Canberra: Australian National University, National Centre for Development Studies, Pacific Policy Paper no. 16.

Ward, Michael (2000) 'Fighting for *ples* in the city: young Highlands men in Port Moresby, Papua New Guinea', in S. Dinnen and A. Ley (eds) *Reflections on Violence in Melanesia*, Sydney: Hawkins Press, 223–38.

Ward, R.G. (1959) 'The population of Fiji', *Geographical Review* 44: 322–41.

—— (1961) 'Internal migration in Fiji', *Journal of the Polynesian Society* 70: 257–71.

—— (1971) 'Internal migration and urbanisation in Papua New Guinea', in M. Ward (ed.) *Population Growth and Socioeconomic Change*, Canberra: New Guinea Research Bulletin no. 42, 81–107.

—— (1973) 'Urbanisation in the Pacific: facts and policies', in R.J. May (ed.) *Priorities in Melanesian Development*, Canberra: Australian National University, and Port Moresby: University of Papua New Guinea, 362–72.

—— (1989) 'Earth's empty quarter? The Pacific islands in a Pacific century', *Geographical Journal* 155: 235–46.

—— (1990) 'Subnational development in Pacific island countries: a dimension forgotten?', *Regional Development Dialogue* 11 (4): 1–17.

—— (1994) 'Managing indigenous land in the South Pacific', *Development Bulletin* 31: 47–9.

—— (1995) 'Land, law and custom: diverging realities in Fiji', in R.G. Ward and E. Kingdon (eds) *Land, Custom and Practice in the South Pacific*, Cambridge: Cambridge University Press, 198–249.

—— (1997) 'Changing forms of communal tenure', in Peter Larmour (ed.) *The Governance of Common Property in the Pacific Region*, Canberra: Australian National University, National Centre for Development Studies, 19–32.

—— (2000) 'Dilemmas in Pacific island land tenure', *Dialogue* 20 (3): 16–22.

—— and Kingdon, E. (eds) (1995) *Land, Custom and Practice in the South Pacific*, Cambridge: Cambridge University Press.

Watts, M.J. (1992) 'Space for everything', *Cultural Anthropology* 7: 115–29.

Wegelin, E. A. (1990) 'New approaches in urban services delivery', *Cities* 7: 244–58.

Weinand, H. and Ward, R.G. (1979) 'Area preferences in Papua New Guinea', *Australian Geographical Studies* 17: 64–75.

Wendt, A. (1977) *Pouliuli*, Auckland: Longman Paul.

—— (1984) *Leaves of the Banyan Tree*, Garden City, NY: Doubleday.

Wessen, A., Hooper, A., Huntsman, J., Prior, I. and Salmond, C. (1992) *Migration and Health in a Small Society: The Case of Tokelau*, Oxford: Oxford University Press.

Western Samoa (1993a) *National Housing Policy Statement of the Government of Western Samoa*, Apia: Government of Western Samoa.

—— (1993b) 'Country paper for Western Samoa', unpublished paper to World Bank Pacific Island Transport and Urban Environment Seminar, East–West Center, Honolulu.

Western Samoa Department of Public Works (1984) *Apia Town Plan 1984*, Apia: Department of Public Works.

White, O. (1965) *Parliament of a Thousand Tribes*, London: Heinemann.

Whitelaw, J.S. (1964) 'Suva: capital of Fiji', *South Pacific Bulletin* 14 (3): 33–7, 54.

—— (1967) 'Gold production and the mining community in Fiji', *New Zealand Geographer* 23: 1–15.

Willis, I. (1974) *Lae: Village and City*, Melbourne: Melbourne University Press.

Winpenny, J. (1994) *Managing Water as an Economic Resource*, London: Routledge.

Wolfers, E. (1975) *Race Relations and Colonial Rule in Papua New Guinea*, Sydney: Australian and New Zealand Book Company.

World Bank (1978) *Papua New Guinea: Its Economic Situation and Prospects for Development*, Washington, DC: World Bank.

—— (1993a) *Towards Environmental Strategies for Cities*, Urban Development Division Strategy Paper, Washington, DC: World Bank.

—— (1993b) *Rapid Urban Environmental Assessment: Lessons from Cities in the Developing World*, vol. 1: *Methodology, Preliminary Findings and Summary of Results*, Washington, DC: World Bank.

—— (1993c) *Towards Higher Growth in the Pacific Island Economies*, Washington, DC: World Bank.

—— (1995a) *Managing Urban Environmental Sanitation Services in Selected Pacific Island Countries*, Washington, DC: World Bank.

—— (1995b) *Papua New Guinea: Delivering Public Services*, Washington, DC: World Bank.

—— (2000a) *Cities, Seas and Storms*, vol. II: *Managing Pacific Towns*, Washington, DC: World Bank.

—— (2000b) *Pacific Regional Strategy*, Washington, DC: World Bank.

World Commission on Environment and Development (1987) *Our Common Future*, New York: United Nations.

Worsley, P. (1964) *The Third World*, London: Weidenfeld.

Wu, D.Y.H. (1982) *The Chinese in Papua New Guinea: 1880–1980*, Hong Kong: Chinese University Press.

Young, R. (2000) 'Gender, mobility and urban place in Fiji: from colonial to post-colonial wanderings', *Asia-Pacific Population Journal* 15 (3): 57–71.

Zimmer, L. (1990) 'Conflict and violence in Gende society: older persons as victims, troublemakers and perpetrators', *Pacific Studies* 13: 205–24.

Zimmer-Tamakoshi, L. (1998) 'Women in town: housewives, homemakers and household managers', in L. Zimmer-Tamakoshi (ed.) *Modern Papua New Guinea*, Kirksville, PA: Thomas Jefferson University Press, 195–210.

Index

For Product Safety Concerns and Information please contact our EU representative GPSR@taylorandfrancis.com
Taylor & Francis Verlag GmbH, Kaufingerstraße 24, 80331 München, Germany

www.ingramcontent.com/pod-product-compliance
Lightning Source LLC
LaVergne TN
LVHW011125300926
849160LV00008B/509
9780415513845